The Transhumanist Movement

Francesco Paolo Adorno

The Transhumanist Movement

palgrave
macmillan

Francesco Paolo Adorno
Dipartimento di Scienze
del Patrimonio Culturale
Università degli Studi di Salerno
Fisciano, Italy

ISBN 978-3-030-82422-8 ISBN 978-3-030-82423-5 (eBook)
https://doi.org/10.1007/978-3-030-82423-5

Cover illustration: Contributor: Kittipong Jirasukhanont/Alamy Stock Photo

This Palgrave Macmillan imprint is published by the registered company Springer Nature Switzerland AG
The registered company address is: Gewerbestrasse 11, 6330 Cham, Switzerland

Contents

1 Introduction 1

2 The Transhumanist Movement 13
2.1 The Word 13
2.2 The People 15
2.3 The Organisations 26
2.4 Declarations 30
2.5 The Definitions of Transhumanism 33
References 37

3 Transhumanism Between Humanism and the Posthuman 41
3.1 Introduction 41
3.2 Humanism 45
3.3 The Theories of the Posthuman 48
3.4 Is Transhumanism a Form of the Posthuman? 52
3.5 Is Transhumanism a Form of Humanism? 55
References 65

4 **Theory and Practice of Transhumanism** 67
4.1 Enhancement 67
4.2 The Aims of Enhancement 74
4.3 What Needs to Be Enhanced? 80
4.4 Body Enhancement 82
4.5 Cognitive Enhancement 85
4.6 Moral Enhancement 91
4.7 The Search for Immortality 101
References 114

5 **Problems of Transhumanism** 121
5.1 Criticism of Enhancement 121
5.2 Therapy vs Enhancement 129
5.3 Medicine and Transhumanism 136
5.4 The Instrumentalisation of the Body 141
5.5 An Anthropology of Disability 146
5.6 Becoming a Cyborg 153
5.7 Legitimacy of Becoming a Cyborg 158
5.8 The Morality of Technology 165
References 170

6 **Transhumanism and Biopolitics** 175
6.1 Introduction 175
6.2 Biopolitics and Transhumanism 179
6.3 Biopolitics as Life Economics 184
6.4 The Economics, Morality and Politics of Enhancement 187
6.5 Wellbeing Between Medicine and Politics 190
6.6 Enhancing to Manage and Govern 194
References 200

Bibliography 203

1

Introduction

As stated in the title, this book deals with the transhumanist movement. It offers the reader not only a description of transhumanism but *also and above all* an assessment of its underlying theory. In order to achieve this, several notional choices have been made.

First, transhumanism is considered as a movement underpinned by the concepts and theories expounded to a greater or a lesser extent by all the authors discussed. Like every choice, this entailed a process of selection and elimination: certain concepts were selected as the primary focus while others were partially or totally excluded. This in turn meant selecting the material to analyse, which was not restricted to concepts and theories but also involved the elimination of some concrete facts. As a result, the map of transhumanism presented in Chapter 2 is far from complete. Transhumanism now encompasses such a spatially and temporally vast array of names, organisations, movements, trends, groups, newspapers, fanzines, publications, conferences and meetings that it would be impossible to take them all into account. This is not intended as an apology but rather as an acknowledgement of responsibility for a choice that is partly justified by the fact that, despite the limitless

F. P. Adorno, *The Transhumanist Movement*,
https://doi.org/10.1007/978-3-030-82423-5_1

material to study or summarise, most of the works on and from transhumanism are boringly repetitive and redundant. Whether it is Californian biohackers working out of garages, Australian bio-artists, Oxford university researchers, Italian neo-futurists or French techno-socialists, the basic ideas are broadly the same.

Second, it is worth recalling how in reality transhumanism acts on two levels: a first level regards the direct enhancement of the human body through the use of all kinds of prostheses that allow for a better control of individual biological processes. The second level of development of transhumanism relates to an indirect enhancement of the human body through a substantially uncontrollable increase in human intelligence which, according to its theorists, will lead to its replacement with a still indefinable form of artificial intelligence. It is argued that the pace of technological progress is so rapid that we will soon enter a phase defined as "singularity" in which this acceleration, originating from weak human capacities, will be such as to transform human nature in an irreversible and fundamentally unpredictable way. The brain, powerful as it is, is limited in the same way that other parts of the human body are, so despite all the creativity it possesses, it seems to have reached its maximum expansion. Notwithstanding this, it has had the ability to initiate a movement of technological progress that seems unstoppable and that, above all, can transcend the biological limits of the human organ, which it still is, creating higher forms of intelligence. Before the end of the twenty-first century, our intelligence will be implemented on a non-biological support billions of billions of times more powerful than the biological support offered by neuronal matter, with consequences that are completely inconceivable at the present stage of our knowledge.

This second-level transhumanism presents a difference, along with an analogy, with first-level transhumanism that explains why we will not deal with it directly. The difference is that, on the one hand, singularity is a largely predictive theory that focuses on the long-term effects of progress in artificial intelligence and that, at least in the terms in which it is presented, are far in the future. On the other hand, the problems posed by the development of artificial intelligence go far beyond what has been discussed in a *specifically* transhumanist context. The analogy is the particularly negative representation of human biology, proposed by both these groups of transhumanists, even though they can be distinguished and isolated. At the bottom of all the branches of transhumanism, there is the belief that human biological processes—the living as such, not to mention natural life—have too many limitations, too many obstacles, that make no sense and that can now be eliminated without any difficulty. Starting from this common diagnosis, "biological" and "singular" transhumanists focus on two different but parallel ways of achieving the same goal which consists of imagining the humanity of the future. That means that the issues discussed are essentially the same, but the means on which to act to produce and think of a posthuman humanity are diverging. For "biological" transhumanists, it is the human body in its diversity and materiality; for "singular" transhumanists, it is the brain that has the evident characteristic of being a biological part of the human body, but it also has the specificity of producing intangible objects such as thoughts, emotions and feelings. The reasons why I prefer to avoid discussing issues related to singularity starting from artificial intelligence are, hopefully, understandable: from the more concretely biological perspective, focusing on singularity would have been redundant with respect to our problems; from the more specifically cognitive perspective, such a focus would have led us to discuss problems that are largely beyond the transhumanist perspective.

With virtually no facts, few names and a great deal of theory, the assessment of a movement cannot be a mere repetition of the almost identical claims its authors make all around the globe. It must not be limited to merely highlighting the implicitly introduced and employed concepts, the more or less implicit justification strategies and the more or less skilfully exploited historical references in order to gain the reader's approval. Instead, it must also shift through all the theoretical material, evaluate it and point out any inconsistencies, issues, logical and theoretical errors, inaccuracies, misstatements and the more or less questionable approximations that it allegedly identifies in its theoretical corpus and, above all, it must consider any resulting practical and theoretical consequences. Despite an equidistance in principle, this book is far from proposing an apology for transhumanism, which the shelves of bookstores and libraries are full of. On the contrary, after a first reading, it will appear as excessively negative and critical of transhumanism. The fact is that the author is a philosopher who is convinced that the role of his discipline is above all to criticise, to problematise and to show the more or less evident implications of concepts, theories and movements to eliminate unreasonable prejudices, while also offering readers some arguments to rationally justify a choice for or against the theses of the transhumanists. Consequently, a point of view was chosen which is this: a comprehensive critique of transhumanism will be developed through the analysis and articulation of a series of arguments that tend, as mentioned, to highlight the shortcomings and criticalities of the topic; if the reader considers that these arguments are valid then his or her rejection of transhumanism will be justified by a solid and substantial reasoning and not only by the dread of transformations and changes, the fear of the new, or by a prejudice favourable to the status quo. If, on the other hand, these arguments are considered weak, inconsistent, inconclusive or useless, the reader will be reinforced in his or her positive conception of transhumanism.

Chapter 4 sets out a detailed overview of the concepts and practices of transhumanism, starting from what appears to be its pivotal concept: enhancement. This is a word used frequently and inaccurately in everyday language, as indicated by the existence of various synonyms, and whose scope indicates such embedded practices that there seems no point in further analysing its meaning. After all, who does not work or act to improve themselves? Who does not seek to better themselves from every point of view? Trivialising enhancement in this way—unintentionally perhaps but certainly without sparing a second thought—obviously means that it can be excluded *ipso facto* from any kind of debate or, worse, delegitimised. What sense would it make to criticise the enhancement proposed by transhumanists if almost all human actions aim at self-improvement? All this would achieve is a paradoxical questioning of human history. Yet things may not be as clear-cut as they first appear, because transhumanist enhancement differs somewhat from humanity's thirst for improvement and gives rise to concerns that have never been raised before. Concentrating on the purpose of human actions, which are always invariably geared towards improving an individual's condition, distracts attention from the means by which these are to be achieved and hinders the proper assessment of their validity because it reduces the gap between the traditional and biotechnological enhancement proposed by transhumanists. The use of biotechnology is legitimised by novel assumptions and could produce revolutionary effects, such as the achievement of immortality. It remains to be seen just how desirable this and the other enhancements proposed and propagated by transhumanism are and whether the enhancement of certain faculties or the creation of new capabilities can be considered truly positive achievements. Indeed, as transhumanism's promises are fulfilled, doubts arise over enhancement in general as a practice involving biotechnological means as well as its specific applications in the search for clear results.

However, it is not only this allegedly extraordinary biotechnological enhancement that is problematic, but also the desire to remove it from procedures of moral legitimation, which should be all the more necessary when their effects appear at first sight so outlandish. Intuitively, apart from the specific arguments used to explain, justify and support the various forms of enhancement, transhumanism employs a specific strategy to legitimise its theories, affirming the full consistency and continuity of its proposals with movements, knowledge, beliefs and concepts traditionally rooted in the theoretical framework of modernity. Basically, transhumanists claim that they are the worthy and orthodox continuation of all the humanists, enlighteners and scientists who have worked for the advancement of humanity. Transhumanism is not an innovative project but one that aims only to accept the consequences of the very movement of modernity through technological progress. Its insistent reference to such notions as progress, humanity and humanism, and even the very name of the movement itself, rightly restores the meaning of this continuity. Transhumanism claims to be nothing other than the orthodox extension of humanism with a view to becoming its crowning achievement: a group of theories and ideas to explain the very premise of the historical self-narration of our world of material progress.

However, the narrative proposed by transhumanism is too smooth, too flat, to be accepted without detailed analysis. Chapter 3 will investigate its alleged continuity with humanism, a claim that transhumanists frequently make in order to take their "rightful" place among the various schools of thought engaging more or less critically with different forms of humanism. This has resulted in a somewhat problematic image of transhumanism: while it may initially seem fallacious to establish a continuity with humanism, in reality there are certainly valid reasons for considering transhumanism as its fulfilment. This conclusion, which is not devoid of ambiguity, can also be attributed to difficulties in defining the key concepts of humanism present in all the authors, thinkers, philosophers and scientists who consider themselves to be humanists. Notwithstanding the uncertainties shrouding the issue, the fact remains that the transhumanists' claim that they are the continuers and heirs of enlightenment and humanism does have a certain legitimacy.

Chapter 5 delves further into the essential principles of transhumanism and assesses its legitimation on the basis of its continuity with traditional practices and theories in order to focus on one of its fundamental features. Transhumanism argues that, as there is no fundamental difference between enhancement practices and medical therapies aiming to restore good health, the rejection of biotechnological enhancement should paradoxically lead to the same attitude being adopted towards medical therapies. Since such a conclusion is literally inconceivable, transhumanists then argue that just as therapeutic interventions are regarded as essential, the same should hold for enhancement interventions. Here too, an excessively linear and superficial rationale fuels the suspicion that such inconsistent and discontinuous reasoning should somehow be exposed, first by making this analogy between the aims of therapy and enhancement less straightforward, and second by pointing out the consequences it might have on people's daily lives. If every therapy is basically an enhancement and if every enhancement has a therapeutic aim, the dissemination and pervasiveness of enhancement could be symptomatic of a view of human life as not simply a succession of illnesses and diseases but as a disease itself; one might then wonder if survival might not lie in pathologising everyday existence.

It seems clear that a line must somehow be drawn to separate health from illness and enhancement from therapy in order to define the limits of normal existence, although this is no simple task. Normality and health are those human conditions that do not need treatment and could certainly be modified through enhancement, although this would differ considerably from therapy. Other arguments legitimising such enhancement interventions would then have to be defined, since they could no longer be justified by therapeutic necessity. The alleged continuity between therapy and enhancement, which is also justified by historical processes that have progressively modified the means and aims of medicine, has an "ideological" impact with deleterious consequences, as is pointed out in the final chapter.

The problematic nature of the transhumanists' "continuist" strategy also emerges clearly when discussing their anthropology. The central figure of transhumanism is, of course, the cyborg: a being that is stronger, more powerful and better than a "normal" human being from every point of view because it has been hybridised with artificial systems that perform the functions of worn out or inadequate organs or replace them, as in the case of RoboCop or the Terminator. The cyborg is an extremely problematic figure as it seems to portray a new human being which, in the eyes of transhumanism's critics, deserves nothing but ruthless condemnation. Transhumanists and their supporters respond to this by stating that person has always been a cyborg. Critics of the cyborgisation of humanity do not accept that the human race has always been fundamentally hybrid, impure and naturally "contaminated" by the environment in which it evolves and by the tools it conceives and creates. The transhumanists once again claim this should not be wondered at and that no justification or legitimation is warranted: the human being becoming a cyborg does not mark a discontinuity with the evolution of its true nature because we are born cyborgs. Denying a discontinuity in the evolution of humanity aims once again to avoid all criticism of the human enhancement that transhumanism supports, by normalising something utterly outlandish. But exactly what is outlandish in the human being that transhumanism proposes to create? What makes it appear so utterly innovative? A first answer to this question can be found in the means that are to be used for this purpose: biotechnologies. While the impact of the biotechnological manipulation of humanity cannot be predicted, there are at least two implications of the analyses conducted in Chapter 5 that deserve to be highlighted.

The first implication is that technology is not itself neutral towards the consequences it produces. The use of new technologies in the field of biology, medicine and elsewhere has the power to change the relationship that people have with themselves, with other people, with their environment and with the world. The problem thus shifts, and it is up to the transhumanists to provide a fuller depiction of the type of world that biotechnologies are likely to design. However, it is clearly not a question of explaining whether or not human nature will be modified, and in what way; for example whether we will have wings or whether devices will be integrated into our bodies to allow us to breathe under water or to survive in the vacuum of space. Instead, what is required of transhumanists is that they explain how everyday activities, however insignificant, will change as a result of humanity becoming a cyborg, and what undesirable but inevitable effects there might be. Rather than on the enhancement of human beings as such, it is on these alterations that the discussion should focus.

A useful example is provided by the world of sport: there are widespread appeals for anti-doping legislation to be withdrawn because performance enhancement is allegedly within every sportsperson's reach and has been in use ever since competitive sport was invented. Enforcing penalties for doping is like trying to empty the ocean with a spoon: futile, costly and counter-productive. Transhumanists put forward the same argument in this context as that used elsewhere: since doping has always existed, no justification is warranted because it would be like justifying the very existence of competitive sport. At the very least, the burden of proof should fall on those who oppose these centuries-old practices. One might object, however, that the intentional or unintentional effects of biotechnological or gene doping make it totally different from "traditional" doping based on corticoids, amphetamines and hormones. The former intervenes directly, definitively and in depth, acting on the athlete at a submolecular level. The transhumanists retort that even if this were the case, the difference between the two doping techniques is quantitative, not qualitative. Clearly, the debate is far from over. One way out of this impasse might be to introduce another aspect that could be defined as practical assessment. Biotechnological doping alters more than just the sportsperson's performance and not only raises issues of fair play but also and above all changes the way in which sport is perceived and played. An Olympic Games 100-m final contested by athletes with physical or chemical prosthetics and by genetically modified or cloned athletes would call into question the current concept of "the Olympic 100-m final". Given the above forms of doping, it is probably the variations *of this concept* and others of the same type and not those concerning doping and its effects that need to be properly evaluated.

The second implication that deserves scrutiny is the unintentional shift in outlook on human nature that the spread of enhancement interventions will inevitably produce. The trivialisation of enhancement, which is nothing more than the trivialisation of improvement therapies, will produce a pathologisation of existence (as will be discussed in Chapter 6) with absolutely unprecedented moral and political consequences. It would be reductive to see transhumanist anthropology as being entirely centred on the cyborg or, rather, it would be a mistake to view the cyborg as the image of an augmented, upgraded human.

Indeed, there is another side to the cyborg, for example in the case of the athlete Oscar Pistorius, who is a fragile, vulnerable, diminished and disabled man. It is because human beings are deficient or impaired that they need enhancement. Under the triumphant optimism of transhumanist technophilia there lies the sad and compassionate image of a man constantly in need of care and therapy, improvements and support without which his existence would be considerably worse than it is. Contrary to popular belief, transhumanism treats people as if they were disabled and in need of improvement therapies throughout their lives. It proposes a generalised therapeutic solution, whose explicit aim is to restore or attribute faculties capable of improving individual wellbeing through increased freedom of movement and greater control over our body. However, the view that humanity is fundamentally inadequate or in a state of virtual deficiency risks triggering a pathologisation of existence: a process in which all the imperfections of the human body become defects to be eliminated, disabilities to be remedied and diseases to be treated.

The moral and political consequences of transhumanist anthropology and of the equivalence between therapies and enhancement are detailed in Chapter 6. This final chapter follows a theoretical argument based on an assumption external to transhumanism in order to abandon the sterile discussions on the possible negative (or positive) effects of enhancement.

These typically prove fruitless and ultimately hinder an overall assessment of transhumanism not only as a movement promoting individualism and competitiveness in the Western world and thereby revolutionising individual lives, but also as a theory producing more concrete and direct effects. The assumption on which transhumanism is here interpreted is that contemporary politics is centred on the biological life of human beings, which has been the mainstay of power since the seventeenth century. This means that the government of individuals, that is all those practices that aim to make individuals perform a certain type of action and adopt specific behaviour patterns with a view to ensuring adherence to a certain vision of morality, is nowadays achieved through life processes.

From the purely moral point of view, transhumanist imperatives pressure individuals into taking responsibility for their own health and activity, establishing a moral code entirely centred on their behaviour and attitudes. Politically, this would result in a deconstruction of traditional political thought, which would then be entirely modelled on biopolitical imperatives that in no way resemble traditional ones. In other words, this diminished and vulnerable conception of the human being would task politics with the management, strengthening and improvement of people's biological life, the consequences of which would be just as important as those previously mentioned with regard to enhancement. Contemporary biopolitics eliminates the gap between the biological processes supporting mere individual existence and the way in which people try to live their lives, inspired by values, concepts and moral and political judgements.

The result achieved by confusing these and reducing the one to the other is the power to manage and govern people's lives through the orientation imposed on the management of life processes. In other words, the life of individuals is governed through the control and management of human biological needs. Biopolitics thus comprises two factors: the governance and the manipulation of human biology, that is the framework within which the creation of individual life must necessarily take place. More precisely, by establishing itself in relation to this biological element, the lifestyle of individuals would be completely managed and controlled by external forces and thus hetero-directed. If transhumanism were internal to this strategy of government, which adopts and updates its working procedures, then our view of these theories and their proposed enhancement would change completely. Transhumanism, which comes across as a theory that promotes and seeks to fulfil even the most repressed and undeclared human desires, would merely supply yet another means of dominating individual lives to those who seek and know how to use it. Beneath the liberation of bodily forces lurks the danger of an even more ruthless and direct control of people through their bodies.

2

The Transhumanist Movement

Transhumanism is not a homogeneous movement, but a multitude of people, organisations, theories and groups that share a relatively small number of theories, views and concepts. Each of these theoretical centres elaborates unanimously shared propositions and tendencies in such an independent and sometimes contradictory way that it becomes virtually impossible to talk about a transhumanist movement: beneath any superficial convergences there is a magmatic activity continually opening up new and different perspectives. This chapter aims to offer a "roadmap" of the movement, the people and the documents underpinning the origins of transhumanism and its continuously developed and re-elaborated theoretical basis.

2.1 The Word

The transhumanist movement, project or whatever else it may be immediately raises a major problem: its very name. What does "transhumanism" mean? In what sense should these two parts, "trans" and "humanism", be understood? What content of "humanism" is somehow

F. P. Adorno, *The Transhumanist Movement*,
https://doi.org/10.1007/978-3-030-82423-5_2

in transition? In what way is humanity or humanism transitioning? What is the destination, if any, of this as yet not fully identified humanity? What does their union indicate?

Unravelling these problems must inevitably entail a lexical reconstruction of the recurrences of the very term "transhumanism" which, far from being a neologism, originates in a very distant past and has a much more tortuous and complex history than we might imagine.

Numerous sources have been cited regarding the origin of the term "transhumanism", although they are not always in agreement. It is generally accepted that the term was coined by Julian Huxley (grandson of Thomas Huxley, the enthusiastic champion of Darwinian theories, and brother of Aldous Huxley, the author of *Brave New World*) in a 1957 text entitled "Transhumanism", in which he states that the human species could, if it wished, transcend itself in its entirety to realise new possibilities of and for its nature (Huxley, 1957).[1] The word signifies "evolutionary humanism", which more accurately defines Huxley's thoughts on the relationship between evolution and humanity. However, despite this useful clarification, Huxley was not the first to use the term. In Canto I of Dante's *Paradiso* the verb *transumanar* is used to translate Paul's description of the ascent to the third heaven in 2 Corinthians 12 (cf. Vita-More, 2019a), later widely present in Teilhard de Chardin. Harrison and Wolyniak hypothesise that Julian Huxley must have been aware of these older recurrences (from Paul of Tarsus to Dante) through the work of Canadian philosopher, jurist and historian W. D. Lighthall who spoke of "Paul's transhumanism" in one of his lectures in 1940.

Olivier Dard and Alexandre Moatti beg to differ, claiming that the term originated in Jean Coutrot's work from 1939 onwards and that Huxley borrowed it (Dard & Moatti, 2016). The hypothesis is plausible as Coutrot and Julian Huxley's brother Aldous were friends, although there remains a doubt: did Coutrot use the term in talking to Aldous who then passed it on to his brother, or did Julian influence Coutrot via his brother?

---

[1] With regard to its recurrence in Huxley's work, Harrison and Wolyniak have stated that the term first really appears in 1951 in a text entitled "Knowledge, Morality and Destiny", the result of two conferences held in Washington on 19 and 20 April 1951 and then published with the same title in the magazine *Psychiatry* (Harrison & Wolyniak, 2015).

At approximately the same time, Teilhard de Chardin coins and uses terms that belong to the same semantic field as transhumanism, but with a completely different meaning and different intentions. Certainly, it is still about imagining a "new" humanity, but his novelty lies in the radicalisation of a traditional vision of humanity produced by progress and evolution. The evolutionary process does not lead to a continuous investigation or an indefinite knowledge and manipulation of the biological characteristics of the human being, but to a transcendence of this dimension that results in the spiritualisation of the person in an unquestionably Christian perspective (Hottois, 2018, 45–48).

2.2 The People

Analysis of the recurrences of the word "transhumanism" provides a mere summary meaning assigned to the movement: history is essentially the story of the improvement of humanity, however this is understood and whatever its procedures and aims may be. Analysing the reflections of philosophers, scientists and theorists who have set out explicitly or implicitly transhumanist theories or have even used the term transhumanism itself facilitates a more accurate indication of the procedures and aims of these improvements. For clarity's sake, we can first divide the latter into three groups and then review the "real" transhumanists who have been joined by figures supporting their views without being explicitly transhumanistic and other researchers from a variety of disciplines who are harshly critical of them.

The first group (the *precursors*) dates back roughly to the late nineteenth century and is made up of philosophers and thinkers inspired by an idea of progress that was substantively shared, both in its theoretical characteristics and in its somewhat limited—given the level of knowledge at that time—but no less envisioned and desired, practical effects.

The second group (the *forerunners*) includes technologists, engineers and scientists from the early twentieth century, who were converging around the desire to create a "new man". This "new man" ideology, mainly presented through scientific discourses of the time and partially deprived of its most evident political connotations, was also disseminated by all early twentieth-century totalitarian regimes. Emblematically, it was then that the possibility of some form of transhumanism was increasingly being taken into consideration, that is the possibility that humanity had the ability and the knowledge to transcend itself and to guide and modify the evolution of the species, which was notably lacking a positive thrust in the desired direction.

The third group (the *proto-transhumanists*) comprises scientists and philosophers from the late twentieth century who, now in possession of the knowledge and technology to modify the species, have been weighing up the moral legitimacy of their use and the limitations that may have to be placed on the genetic manipulation of individuals on the one hand, and the possibilities opening up for further technologies to manipulate the biological substratum of individuals on the other hand.

All these scientists, researchers, philosophers and technologists (whether precursors or forerunners of transhumanism or proto-transhumanists) share a number of views: they propose an evolutionary conception of human nature; they regard freedom of research and the critical use of reason as values to be promoted; they consider technology and science as tools at the service of the material enhancement of the human being essentially through biological modification; and they embrace a biologically connoted philosophy of history that is more or less consciously based on the history of evolution.

From this point on, however, matters become extremely complicated because the prominence that transhumanist propositions and their prospective implementation assume gives rise to a somewhat heated debate between transhumanists and their dyed-in-the-wool *opponents*, the so-called bioconservatives or bioluddites The debate also sees the involvement of a group of transhumanist *supporters* generally referred to as bioliberals—philosophers, bioethicists and scientists sympathetic to transhumanist ideas while not belonging to the movement itself—whereas bioconservatives are fundamentally hostile to the biological interventionism of transhumanists, whose theories they contest.

Apart from questions of historical philology, identifying the *precursors* of transhumanism encounters two serious difficulties, in addition to the extremely problematic use of this historical category. First, it was the transhumanists who identified their own precursors or regarded themselves as the heirs of several fairly different theoretical traditions, thus establishing a family tree whose members would most likely disagree with such an affiliation. James Hughes, one of the first secretaries of the World Transhumanist Association (WTA) and the theorist of a form of transhumanism that is less libertarian and more attentive to the problems of distributive justice, claims that transhumanism is the convergence of two tendencies characterising humanity across the millennia: on the one hand, its desire to transcend its own limits, something that religions have traditionally sought to achieve by spiritual and cultural means; on the other hand, the progress of science and technology which, while rather limited until the late twentieth century, has recently experienced such an increase that humanity has been able to concretely and materially transcend its limits (on Hughes see Agar, 2010, 150–177; Deprez, 2019). The turning point in this history came with the Italian Renaissance and the Enlightenment, above all with the renewal of scientific method theorised by Bacon, who became the true "guardian" of transhumanism. New science, liberation from religion and freedom of thought merged into the cultural movement that culminated in the French Revolution and assumed a markedly militant aspect thereafter. Hughes sees transhumanists as the heirs of a series of thinkers from Benjamin Franklin to Condorcet and Godwin, bearers of the values of Western rationalism who also seem to have fully espoused the aims of transhumanism because they assumed and imagined that progress could and should aim at significantly lengthening life (Hughes, 2004, 157), and obviously Darwin as well. The closer we get to the twenty-first century, the greater the capability of scientific discoveries and technological inventions to achieve what until then had been considered the figments of a fervid imagination, and the more numerous the precursors identified by transhumanism. Hughes certainly does not fail to include science-fiction writers, such as Isaac Asimov, Sebastian Lem, Alfred E. Van Vogt or Arthur C. Clarke, all of whom understood that new technologies had finally developed the capability to build robots, cyborgs and other forms of humanity, nor does he forget such eminent scientists as John B. S. Haldane, John D. Bernal, Herman Muller and Julian Huxley, who predicted the advent of a new, self-fashioned species enabled by advances in biomedicine and biology. Finally, from Robert Ettinger to Jonathan Glover, transhumanist ideas enter the scientific and theoretical debate, no longer as speculations on a world to come or as astonishing and still uncertain innovations, but as elements that structure people's everyday lives.

This view is substantively accepted and taken up by Bostrom who adds some additional and creatively justified references. Among the humanists of the Italian Renaissance of whom Bostrom considers himself the heir, he explicitly mentions Giovanni Pico della Mirandola because he theorised the substantial indeterminacy of the nature of man. Unlike the opponents of transhumanism who justify their position on the grounds that transhumanist paradigms would illegitimately alter the essence of the human being, Pico della Mirandola saw the greatness of man in his not having a nature and in being able to assume any nature from a beast's to an angel's. In addition to the eighteenth-century philosophers mentioned by Hughes, Bostrom also includes Newton, Hobbes, Locke and Kant because they underpin humanist rationalism, advocating a critical use of reason and the progress of science, which are simultaneously factors of moral and material improvement (Bostrom, 2005a, 2–3).

Max More cites the same philosophers, scientists and engineers among the precursors and proto-transhumanists, but he more vehemently supports the view that transhumanism is nothing more than the fulfilment of the ideals of humanism and, above all, of the Enlightenment, explicitly counterposing it to the post-structuralist and postmodernist schools of thought that break down into forms of rational and moral relativism (More, 2013a, 10).

Transhumanist texts also reflect the ideas of Descartes and Locke. Descartes appears to have anticipated the possibility of mind uploading and his dualism assures the likelihood of this practice, while Locke underpins mental functionalism, which attributes little importance to the physical means for implementing mental data, the production of which does not require a specific biological medium. Opinions differ somewhat with regard to Nietzsche, who intuitively could or should be a theoretical reference for these philosophers. According to Bostrom, Nietzsche has nothing to do with his posthumanism (Bostrom, 2005a, 4), or in any case less than Mill (he believes that Nietzsche would certainly have been pleased, but he is not sure Mill would have agreed), while for another important figure in this movement, Max More, the German philosopher should certainly be among the inspirers of transhumanism.[2] The problematic aspect of this family tree produced by the transhumanists themselves is the presence of eminent representatives of humanism and the Enlightenment among the movement's precursors. It is certainly not surprising that ever since the beginning of the twentieth century there has been, if not the knowledge, at least the justifiable suspicion that humanity has been consciously striving towards the transformation of individuals and the species as a whole. However, that Bacon, Condorcet or eighteenth-century scientists might be named among the precursors of transhumanism is quite a leap and presupposes a reduction of their thought to these elements taken individually and devoid of their context.

The second difficulty arising from this family tree is broader in scope and concerns the legitimacy of transhumanist theory and practice. Claiming to be the heirs of a tradition of thought that coincides with the producers of the very project of modernity has an extremely precise meaning, as it allows any doubts on transhumanism itself to be nipped in the bud. The claim that transhumanism is but the continuation of Renaissance humanism, as its theorists are the heirs of Bacon, Descartes and Condorcet, leaves critics facing two equally problematic alternatives: to refute the transhumanist project by inevitably rejecting the ideas of those who are said to be its inspirers, or to accept that transhumanism must be permissible in that it is the extreme and legitimate consequence of the project of modernity. The rather uncomfortable way out of this predicament lies in demonstrating that transhumanism is a perversion of modern ideals and practices, some features of which have

[2] Against Bostrom's views, cf. More (2010, 1–4). An issue of the *Journal of Evolution and Technology* is devoted entirely to the analysis of the relationship between the Nietzschean superman and transhumanism. To reconcile the two opposing positions, it could be said that, without Nietzsche, a philosophy of the *Übermensch* would probably not have been possible, but at the same time, his idea of the man of the future is quite different and even opposite to that of the posthumanists. However, Nietzsche was the subject of a heated debate on the presence of a view compatible with transhumanism in his work: cf. Sorgner (2009). A long and comprehensive discussion of Nietzsche's possible enlistment by transhumanists is conducted by Hauskeller, who denies that this is legitimate (Hauskeller, 2016, 75–84).

been selected but whose essential parts have been abandoned. The examination and elaboration of this topic will be discussed at length in the next chapter, focusing on the relationship between humanism and transhumanism. For the time being, we will focus on other groups of thinkers, scientists and philosophers who have somehow cultivated the garden of transhumanism: the forerunners and the proto-transhumanists.

There seems to be no doubt that the forerunners of transhumanism in the English- and French-speaking world present theories and ideas that are well rooted in the spirit of the time and viewed favourably beyond a particular *koinè* (Coenen, 2014). The earliest examples of transhumanism appeared in a context where the totalitarian regimes emerging in the aftermath of the First World War underlined the need to "create" a new man. From Mussolini to Trotsky, the stress fell repeatedly on the need to renew humanity in accordance with their respective ideologies. Fascism's new man would be fundamentally different from that of Soviet Communism: above all, an anti-bourgeois, nonconformist individual with moral, social, political and economic qualities completely opposed to the now hegemonic bourgeois figure of contemporary European societies, described by Werner Sombart, parodied by Robert Musil and criticised by Thomas Mann. The new man that Fascism intended to create had very different moral and political qualities from those of the European bourgeoisie, acquired not through biological manipulation but through a different "upbringing".

While it cannot be denied that Fascism's imperative to build a new man was also shared by Trotsky, considerable differences lie not only in the ends but also in Communism's essentially biological means to achieve those ends (Hauskeller, 2016, 28). Trotsky's theories are surprisingly compatible with transhumanism and support the idea that the renewal of man, however this is to be achieved, is a widely felt need that is much older than transhumanist views. Julian Huxley was certainly not indifferent to this general feeling, given his closeness to two other forerunners of transhumanism, Haldane and Bernal, both members of the Communist Party of Great Britain (CPGB) (Tirosh-Samuelson, 2011a, 63). While Huxley was one of the first to use the term transhumanism in a sense very close to that of contemporary transhumanists, it cannot be said that he was waiting for the name in order to grasp the concept. It is true that, as Hughes and Bostrom are at pains to point out, there are numerous myths about human beings eager to transcend the limitations of their immanence, but it is also true that questions on the future of humanity, the formative role of science and technology and the evaluation of the consequences of progress are particularly pressing at this time. Thus, Huxley seems to be not only the spokesman of an era but, in view of the meaning he confers on the term transhumanism, a true forerunner of the theories of contemporary transhumanists. Underpinning Huxley's theories is a central theme of present-day transhumanism, namely that evolution is now in the hands of human beings, who are no longer mere objects but actors in their own right. Despite this and other very similar views, there are at least two elements that complicate this filiation with transhumanism.

First, Huxley sees it as a question of achieving the human being's potential and fulfilling its nature, not of going beyond the limits imposed on it by biology, which seems to be the aim of transhumanism (Goffi, 2015). However, there is no unanimous, transhumanist view on the matter and, indeed, it is difficult to quantify the extent to which supporters of transhumanism concur on these theories: for some, it is a question of striving towards an indefinite perfection of human nature while remaining within its limits but, for others, the aim of transhumanism is the creation of a new species.

Second, the enhancements that Huxley sees as absolutely necessary are to be applied to the population as a whole, not on an individual basis, so as not to reinforce inequalities, but rather as an attempt to eliminate them. Underpinning his transhumanism, also known as "evolutionary humanism", is a social holism, rather than the exasperated individualism that seems to connote transhumanism (Bashford, 2013). In order to emphasise the distance from the themes of contemporary transhumanism, or at least from those espoused by Max More and the Californian group but not from Hughes's "democratic transhumanism", we may notice that Huxley puts forward the creation of an appropriate social environment inspired by ideals of beauty, which is an end in itself (Huxley, 1957, 76). This does not detract from the fact that his "evolutionary humanism" envisions an enhancement that will lead to a "new kind of existence".

In *Daedalus or Science and the Future*, Haldane presents the foreseeable effects of scientific progress (especially in biology, which he considers the fundamental science of the centuries to come) in a utopia scientifically based on a democratic use of eugenics (Haldane, 1923). According to Haldane, advances in biology will have a positive effect on medicine, which will not only provide the means for an increasingly better life but will also lead to the creation of a race of superhumans through the application of eugenic principles. The envisaged progress is practically identical to that of any other utopia or attempt at historical prediction: the development of human faculties, the abolition of disease and the sweetening of death so that it resembles sleep rather than a traumatic event (Tirosh-Samuelson, 2011a, 69), but also the elimination of world hunger and the development of new forms of artificial procreation.

Bernal's theories also go in this direction. In *The World, The Flesh and the Devil: An Enquiry into the Future of the Three Enemies of the Rational Soul*, he outlines a scientific utopia based on the latest discoveries and inventions and speculates on the nature of the humanity of the future, accurately anticipating some of transhumanism's central issues, such as mind uploading (Bernal, 1969 [1929], 41). He claims it will be possible to insert the brain into a cylinder and connect it to a new body with new senses, thus creating a new person. Human beings will acquire new faculties as a result of the new operating procedures of a body equipped with "television apparatus, tele-acoustic and tele-chemical organs and tele-sensory organs of the nature of touch for determining all forms of texture bodies" (ibid.). This new man is merely the foreseeable consequence of evolution (Tirosh-Samuelson, 2011a, 74–75).

While Haldane and Bernal lie between the envisioned enhancement of humanity and a techno-scientific utopia, in the mid-twentieth century biological manipulation became a reality, certainly in need of perfection but nevertheless utterly feasible. The transhumanists themselves drew up a list of those that they considered *proto-transhumanists*: the theorists of artificial intelligence such as John von Neumann, Marvin Minsky, Hans Moravec and Vernor Vinge; Eric Drexler, author of the first text on nanotechnologies, anticipating the possibilities of intervening in the molecular substratum of living beings (Drexler, 1986); Robert Ettinger, who published *Prospect of Immortality* in 1962 which sets out the idea of cryonic suspension; and Jonathan Glover, one of the first to investigate the social, economic and political consequences of the new power enabling the genetic selection of individuals in *What Sort of People Should There Be?* in 1984. Unlike their predecessors, the latter three authors not only have a solid scientific basis for identifying lines of development in biology that leave little room for flights of fancy, but they also question the moral legitimacy of their use. The utopian element is still present but limited to an investigation of the changes that these new technologies (molecular nanotechnology, cryonics and genetic engineering) will produce, starting from an assessment of their already ascertained potential.

The first transhumanist proper is undeniably Fereidoun M. Esfandiary, better known as "FM-2030", although it would be more appropriate to consider him a proto-transhumanist. When called upon to define the characteristics of future human existence, Huxley limits himself to saying that it will be "as different from ours as ours is from that of Pekin man" (Huxley, 1957), whereas Esfandiary proposes a more precise vision of the "transitional man" (Esfandiary, 1989) which includes colourful though actually accomplished features, such as the intensive use of telecommunications, a globetrotting lifestyle, a cosmopolitan outlook and totally liberated sexual behaviour.[3] More concretely, the transitional man would acquire greater control over himself through emancipation from the body, the source of all human weaknesses. This greater autonomy would serve a predominantly recreational purpose, contrary to what Huxley advocates: it should broaden the horizons of human pleasure and constantly provide new and diverse experiences (Goffi, 2015) and project man into a context where technology rules and can be exploited to the best possible advantage. Esfandiary's best-known book

[3] As also pointed out by Sorgner (2009). Bostrom correctly observes that it may well be true that "the signs that FM saw as indicative of transhuman status included prostheses, plastic surgery, intensive use of telecommunications, a cosmopolitan outlook and a globetrotting lifestyle, androgyny, mediated reproduction (such as in vitro fertilisation), absence of religious belief, and a rejection of traditional family values", but it is quite a stretch to then define a "posthuman" because "it was never satisfactorily explained why somebody who, say, rejects family values, has a nose job, and spends a lot of time on jet planes is in closer proximity to posthumanity than the rest of us" (Bostrom, 2005a, 13–14).

is *Are You a Transhuman?* (1989), but his ideas are more fully developed in the trilogy *Optimism One*, *Up-Wingers: A Futurist Manifesto* and *Telespheres* (1977a[1970], 1978[1970], 1977b), in which he sets out a philosophy of optimism based on two evolutionary advances: on the one hand, as we overcome the limitations of our mortal bodies we can extend our life indefinitely, and, on the other hand, we are on the verge of accessing the infinite resources and possibilities hidden in the universe (1977b, 170). The means to bring about this progress is obviously technology, which will ensure freedom from the biological limitations that prevent all human beings from achieving their full potential. Since "the body has been our greatest inhibition, our most serious obstacle to an evolution that must lead us higher" (Esfandiary, 1977a, 166), the task before us is precisely the need to remake the human body and transform it into something beautiful, varied, fluid and durable (ibid., 167–168).

However, as stated above, the transhumanist galaxy is extremely varied, and it would serve no purpose to list the names of all adherents to transhumanist ideals. In addition to the authors already mentioned, the contributors to the three anthologies *The Transhumanist Reader* (More & Vita-More, 2013), *H+: Transhumanism and its Critics* (Hansell & Grassie, 2011) and the more recent *Transhumanism Handbook* (2019a) edited by N. Lee, can also be considered transhumanists in their own right.

At this point we are faced with a fairly serious problem regarding our definition of the transhumanist movement. How can we identify authors who may not accept being regarded as transhumanists but can be said to be *supporters* in that they embrace views compatible with those of transhumanism (but not necessarily analogous to them), even though they are not affiliated to any transhumanist organisation? Indeed, apart from the figures already mentioned, who have actively participated in the formation and the consolidation of the transhumanist movement, there are other researchers, the bioliberals mentioned above, from a wide range of disciplines, who subscribe to views and theories that are absolutely consistent with those of transhumanism (although not identical to them) but politely reject this label, which they are beginning to find rather awkward, such as Aubrey de Grey, a transhumanist of the first hour. Among the bioliberals, mention should be made of the more famous technophiles Allen Buchanan, Dan Brock, Norman Daniels, Daniel Wikler, Nicholas Agar, John Harris, Arthur Caplan and Gregory Stock, who enthusiastically supplied moral legitimation to the biotechnologies of enhancement. Even closer to the transhumanist position is Julian Savulescu, director of The Oxford Uehiro Center for Practical Ethics and author of numerous articles, some in collaboration with Ingmar Persson and others with Nick Bostrom or Anders Sandberg, in which transhumanist views are defended to the hilt.

However, it is no simple task to explain how Bostrom's militant technophilia (although he did take a more circumspect position on artificial intelligence, one of transhumanism's favourite themes) differs from the path followed by a liberal like Nicholas Agar, who set out to defend enhancement and even wrote a lengthy apology of liberal eugenics but now rejects radical enhancement in favour of a truly more moderate one (Agar, 2014). Indeed, the differences between them do not seem as marked as they might first appear. It has been said, for example, that while true transhumanists adopt enthusiastic technophile positions aiming to legitimise the creation of a new species, that is that biotechnologies must literally create new capabilities and thus ultimately a human being 2.0, bioliberals are more cautious and believe that the use of biotechnologies must push human faculties to the limit of their potential but not beyond, because of the currently unforeseeable risks that their indiscriminate use might pose to the continuation of life on Earth. In actual fact, the issues are more nuanced: More has explicitly stated that transhumanists have not set themselves the goal of creating a new species; and Harris, a fervent bioliberal, does not seem to pose the problem of the limits of human enhancement. The extent to which transhumanists' prudence and bioliberals' enthusiasm might be a tactical ploy remains to be seen. However, it is worth mentioning Buchanan's proposal that the so-called bioliberals, while taking care not to claim explicit affiliation to the various transhumanist organisations but nevertheless adopt their views and develop their theories in the same direction as them, should be considered as being "anti-anti-enhancement" and not as active supporters of the benefits of enhancement.

On the other hand, it seems easier to identify the *opponents* of transhumanism, whether they be *bioconservatives* or *bioluddites*. The most frequently recurring names are Leon Kass, Francis Fukuyama, Michael Sandel and Jurgen Habermas. James Hughes divides the bioconservatives into two groups, the right-wing and left-wing bioluddites, which differ essentially in the type of criticism they make of the free, unchecked, indiscriminate and unregulated use of biotechnologies for enhancement. Right-wing bioconservatives such as Kass and Fukuyama adopt arguments based on religious positions, and more generally on a sacralisation of nature, which should be altered as little as possible. Left-wing bioconservatives such as Michael Sandel, Jurgen Habermas or Langdon Winner, on the other hand, are more attentive to the undesirable risks and changes for humanity incurred through an indiscriminate use of biotechnology for the purposes of enhancement. They feel that enhancement practices need to be monitored, not to limit their use but to garner an understanding of their general effects, thereby avoiding the adoption and use of technologies that we do not fully control. In short, right-wing bioluddites are in favour of limiting and banning certain specific practices in principle, while left-wing bioconservatives advocate gaining a better understanding of the effects and consequences of enhancement.

2.3 The Organisations

Transhumanism is not only a set of theories about the nature, characteristics and potential evolution of humanity, it is also a movement made up of a series of organisations that have systemised and propagated these ideas over time (cf. Marzocco, 2019, 36–81).

The transhumanist movement has passed through three phases. The first phase was the creation of the Extropy Institute by a group that included Max More. As early as 1986 Max More had joined Alcor, a cryogenic company of which he is currently CEO and with which he helped establish (along with Michael Price, Garret Smyth and Luigi Warren) the first European cryonics organisation, Mizar Limited (later Alcor UK). In 1988, he moved permanently to California where he changed his name from O'Connor to More and set up *Extropy*, a journal presenting the key concepts of extropy[4] and the first draft of transhumanism, with the help of another young researcher, T. O. Morrow (previously called Tom Bell). This was followed in 1992 by the founding of the Extropy Institute, whose activity intensified in 1994 with the organisation of a series of conferences on themes that were to become central to transhumanism (Vita-More, 2019a).

Extropianism's ideas quickly spread far and wide thanks in part to the important contribution made by Max More's wife, Natasha Vita-More, to the Extropy Institute's growth. Previously the companion of Fereidoun M. Esfandiary, she developed transhumanist themes in the field of art and cultural projects. A victim of its own success, the Extropianism movement was dissolved in 2006 but its deep-rootedness in California (one of the fundamental characteristics of transhumanism and its technological developments) continued. Silicon Valley is effectively the cradle in which attempts are carried out to translate transhumanist visions into practical reality. The relationships with successive generations of engineers, scientists and technologists are absolutely organic in both practical and theoretical terms, as can be clearly seen in the extent to which the US high-tech industry and explicitly transhumanist organisations and foundations are entwined. The figure of Ray Kurzweil is emblematic of this union of daring speculation, theoretical predictions on the future of humanity and the promotion of scientific research in well-defined fields like artificial intelligence or biochemistry (Damour & Doat, 2018). Kurzweil too, like all the other transhumanists, recognises that the human body, in particular the brain, suffers from important limitations: it is too slow and with a reduced calculation capacity, even taking into account its parallel functioning; it is too fragile, and in need of too many costly treatments. However limited, the brain has at least been able to initiate an enhancement process that will lead its creations

[4] Extropy is "the extent of a living or organisational system's intelligence, functional order, vitality, energy, life, experience, and capacity and drive for improvement and growth. Extropy is not a rigorously defined technical term in philosophy or science; in a metaphorical sense, it simply expresses the opposite of entropy". These early lines suggest a never denied interest in science fiction as one of the main sources of inspiration for transhumanism: the first usage of the term "extropy" to indicate a potential transhuman destiny for humanity is in fact found in a 1983 novel by Diane Duane, *The Wounded Sky*, from which the first season of the television series *Star Trek: The Next Generation*, episode "Where No One Has Gone Before", was adapted.

to replace it more or less quickly. Kurzweil's diagnosis is that "by the end of this century, the nonbiological portion of our intelligence will be trillions of trillions of times more powerful than unaided human intelligence" (Kurzweil, 2005, 25). In other words, human beings are acting for their disappearance and their replacement with artificial intelligences which should take place by the end of the twenty-first century. From a strictly mathematical perspective, the growth rates will still be finite but so extreme that the changes they bring about will appear to rupture the fabric of human history. That, at least, will be the perspective of unenhanced biological humanity. Singularity will represent the culmination of the merger of our biological thinking and existence with our technology, resulting in a world that is still human but that transcends our biological roots. There will be no distinction, post-Singularity, between human and machine or between physical and virtual reality. However, this will not lead to the disappearance of the human species, but to a profound transformation that will leave only one of its characteristics intact, brought to a level of improvement never reached by the human species: the research "to extend its physical and mental reach beyond current limitations" (Kurzweil, 2005, 25).

In 1998 Nick Bostrom and David Pearce founded the WTA in Europe. The WTA brings together numerous transhumanist associations from all over the world, focusing primarily on the theoretical aspects of the project, as well as researchers such as the sociologist James Hughes, Max More himself, Anders Sandberg and Kathryn Aegis, feminist and activist for the rights of handicapped people. Transhumanism has now spread all over the world but is deeply rooted in Oxford, where Bostrom is a teacher and director of The Future of Humanity Institute, and Julian Savulescu is the director not only of The Oxford Uehiro Centre for Practical Ethics but also of The Oxford Centre for Neuroethics, which hosts yearly conferences on the favourite themes of transhumanism.

The third phase of the movement recently saw the WTA change its name to Humanity +, which is still active today and preserves its previous federal structure. Mention should also be made of the Institute for Ethics and Emerging Technologies (IEET) founded by Nick Bostrom, Mark Walker and James Hughes which brought the leading theorists of transhumanism together under the banner of techno-progressivism, an idea defined as an ideological stance with roots in Enlightenment thought which focuses on how human flourishing is advanced by the convergence of technological progress and democratic social change. Technoprogressives argue that technological innovations can be profoundly empowering and emancipatory when they are democratically and transparently regulated for safety and efficacy, and then made universally and equitably available.

The transhumanists also became involved in politics: in 1992 Natasha Vita-More was elected as a council member for the 28th Senatorial District of Los Angeles and, after several attempts, they even put up a US presidential candidate, Zoltan Istvan Gyurko, explicitly inspired by their programme. These early experiences met with varying degrees of success and culminated in the founding of the first US Transhumanist Party, whose leader since 2016 has been Gennady Stolyarov II (Lee, 2019b, 16).

In addition to the adoption of the same theories, another element of continuity shared by these different organisations is the constant presence among their founders and members of numerous well-known personalities acting as catalysts to attract new researchers and other theorists of transhumanism. These include Natasha Vita-More, president of the Extropy Institute and now director of Humanity+, whose members include Max More who is also CEO of Alcor and among the fellows of the IEET. Anders Sandberg has followed the WTA's transition into Humanity +, as have Aubrey De Grey (also a fellow of the IEET and co-founder of Strategies for Engineered Negligible Senescence (SENS)—Research Foundation), Nick Bostrom and James Hughes.

Other organisations were born out of the desire to put the research that inspired transhumanism into practice, although they were not directly involved in the promotion of the movement or its general project but focused on specific aspects of transhumanism. One particular example is the Foresight Institute, whose website explains that it was founded on a vision set forth by Foresight's co-founder K. Eric Drexler in *Engines of Creation* (*Engines*), published in the spring of 1986. Foresight was founded coincident with the publication of *Engines* to research and advance the theme of the book—a web of emerging revolutionary technologies that will present opportunities to vastly improve the human condition, and a web of strategies to secure those benefits while avoiding the problems that could come from careless or malicious misuse of those technologies, or from unintended consequences.

Finally, the SENS Research Foundation, founded in 2009 by De Grey and Michael Kope, conducts research seeking to determine the causes of ageing and death. Immortality is the main goal of transhumanism and the real threshold that needs to be crossed in order to create a posthumanity. A network of research institutes with the same aims has been established around SENS, including the Methuselah Foundation.

2.4 Declarations

The principles of transhumanism were first presented in a short article by Max More entitled "Principles of Extropy" and published in 2003, which sets out and organises a series of considerations dating as far back as the early 1990s. More explains that extropianism is centred on the individual, who is free to decide their own life plan and to improve their existence in any way they choose (self-transformation and self-direction). The most appropriate means of achieving this goal are technological innovations applied "creatively and courageously to transcend 'natural' but harmful, confining qualities derived from our biological heritage, culture, and environment" (*intelligent technology*). Also required is a practically boundless belief backed up by *practical optimism* in the enhancive power of *rational thinking*, whose constructive but critical work keeps false beliefs in unanalysed dogmas at a distance and eliminates the tendency to fall into paralysing forms of pessimism. This rational optimism translates into the conviction that humanity cannot but progress and improve its life conditions indefinitely (*perpetual progress*). Accompanying these positive transformations there must be an *open society*: Extropy means supporting social orders that foster freedom of communication, freedom of action, experimentation, innovation, questioning, and learning. Opposing authoritarian social control and unnecessary hierarchy and favouring the rule of law and decentralisation of power and responsibility. Preferring bargaining over battling, exchange over extortion, and communication over compulsion. Openness to improvement rather than a static utopia.

The first transhumanist declaration proper was issued in 1998 and, in accordance with More's principles of extropy, its stated intention was to promote research in three specific fields: ageing and immortality, artificial intelligence, and the conquest of space. The last of these would soon be relegated to a secondary position while artificial intelligence became the object of in-depth research stretching far beyond the transhumanist movement which, as Bostrom stated, adopted a more cautious approach to its possible evolution and any potential dangers. This led to a sort of theoretical skimming that effectively focused the efforts of transhumanism on improving the human condition by intervening in its biological heritage, which may now be said to be the *core business* of transhumanist theories (O'Connell, 2017). Having abandoned speculation on the possibility of transferring the human race to other planets and lowered its expectations of enhancement through the development of artificial intelligence, transhumanism now focuses essentially on the prospects achievable through direct intervention in human biology. The main aim of transhumanism is now the definition and the moral, political and social legitimation of its positive impact on the human condition—the elimination of the frailty of the human body (disease, suffering, disability, limitations and death) by means of genetic modification, biological manipulation and pharmacological and physical prosthetics. The extensive application of technology can only produce positive results as it is a fundamental factor of emancipation from the limitations of the human condition. Furthermore, it is a key tool in the promotion of individual freedom which, despite an evident reluctance to develop its political implications, is regarded as the priority value and is limited only by respect for the freedom of others.

The 2012 version of the *Manifesto* promotes individualism and technophilia with the same enthusiasm but with a greater awareness. The positive function of technology is somehow counterbalanced by a recognition of its ambivalence. Transhumanism concedes that technology has an undeniable potential to endanger life in general and human beings in particular. What is more, its adherence to the concept of ultraliberalism diminishes between the first and the second declaration, although reference to the importance and priority of individual choices remains strong.

As stated above, transhumanism has split, at least from an institutional point of view, into two entities: Humanity+ and the IEET. While the former has completely embraced the means, theories and aims of WTA transhumanism and is effectively a theoretical continuation of Californian technophile, scientistic and libertarian transhumanism (Vita-More, 2019a), the latter issued a declaration in 2014 stressing a more balanced view of the benefits of technology than early transhumanist declarations. While taking up the central idea of extropianism and transhumanism (namely that the products of reason, technological innovations and scientific discoveries are fundamental to the progress of humanity), technoprogressivism highlights the risks involved in the use of new technologies and instead recommends proceeding with some caution. The clear reference to the ideals of the Enlightenment is not limited to recalling the function that human rationality performs in material progress, but it also points out that such progress was conceived as a factor promoting freedom and social justice (Coeurnelle & Roux, 2016). Unlike early transhumanist declarations, the manifesto of technoprogressivism dwells at length on the need to define and create the political and economic conditions necessary to ensure a fair and equitable distribution of resources and technological innovations with a view to reducing inequalities. It states that not only is it totally unacceptable for technological progress to be discriminatory but also that its aim is to improve conditions for all humanity, which cannot be achieved outside a democratic framework.

Despite a global continuity of ideas, the shift from extropianism to transhumanism and then to technoprogressivism initially reduced the liberal anarchism present in Max More's theories (Hughes, 2004, 164–169) and then raised awareness of the dangers and the damage that an over-enthusiastic technophilia could cause. Transhumanism progressively adopted more consensual political views resulting in positions differing from those presented, for example, in Max More's 1991 article "Orders without Orderers" which was based on Ayn Rand's ultraliberal ethics,

Friedrich Hayek's economic liberalism or David Friedman's anarcho-capitalism. The technoprogressivism of the IEET is based on the idea that transhumanism is a revival of the ideals of the Enlightenment which saw material progress as the condition and effect of social and moral progress: political conditions aimed at the affirmation of social equity, as well as the consolidation and enlargement of a democratic regime, were necessary conditions for a broader and better individual freedom. In this framework, economics became a sort of useful political tool for increasing global wellbeing in very specific directions: justice takes priority over freedom or at least is an essential condition for it.

2.5 The Definitions of Transhumanism

Apart from the questions regarding its position in the cultural panorama at the beginning of the new millennium and the evolution of its main concepts, transhumanism is by its very nature extremely difficult to define because it is a movement and a theory centred on a transition, a transformation, a dynamic. Transhumanism supports, asserts and strives so that the human being, in the form that we are partially able to know or recognise when we reflect upon ourselves, is radically transformed or perhaps even disappears as a result of technological and scientific development. If the fate of all living species is sooner or later to perish under the pressure of evolutionary processes, then the end of the human race (at least in its present form) is desired and brought closer not only by uncontrollable biological dynamics but by humanity itself or at least by transhumanists. The human race could thus give way to a species whose form we are apparently already beginning to discern and which could be completely, though not necessarily, different from today's human beings.

Discussing transhumanism is, therefore, extremely problematic if only for this reason alone. The first real difficulty stems from the fact that the form of this new species to come or, more unassumingly, of the changes that might modify some features of the human race, are so indefinite and indefinable that visions of this human future are identified by a panoply of prefixes: transhumanists, hyperhumanists, posthumanists, not to mention the borganists (Sandberg, n.d.) or metahumanists

(Stock, 1993). While all these theories offer blurred and very different images of the men and women of the future (although something resembling sexual gender is conceivable), they nevertheless agree that they will still be "better" (whatever that means) than the human beings of today. Starting from the rather trite observation that humanity is undergoing a transformation, transhumanism aims to bestow moral, political and social legitimacy on this change by familiarising us with the idea that future generations will have very different characteristics from those of past generations as a result of the more or less voluntary application of new technologies to the human body. The reference to an as yet inadequately defined posthumanity might generate a certain confusion with the posthuman cultural movement, although this can be easily overcome, at least for the time being, by referring to transhumanist declarations. Although extremely concise, these declarations set out the clear and precise principles that every transhumanist explicitly and convincingly embraces, which are very different from, not to say opposed to, those set out by the extremely heterogeneous posthuman movement.

Apart from the above issues, defining transhumanism presents no further difficulties because its supporters have dealt with it exceedingly well: not only have they set out well-organised, detailed and precise definitions in their writings, but they have also presented them clearly and schematically in their declarations and manifestos. Regardless of the desire to make transhumanism acceptable, which entails toning down statements that are overly apodictic and fairly untenable either because they are too optimistic or because they go against the mainstream, it is nevertheless clear that the aims of transhumanism have remained unchanged and easily identifiable over the years.

Transhumanism is "the belief in overcoming human limitations through reason, science, and technology" (Young, 2006) or, quite simply, "the thesis that we can and ought to use technology to alter and improve human biology" (Walker, 2014). Walker is also very effective in clarifying that transhumanism implicitly assumes that "intelligence, moods, longevity, and virtues each have deep roots in our biology", and hence "by altering biology, transhumanists propose to improve human nature to the point of creating a new genus: posthumans" (Walker, 2014). These posthuman beings would constitute a new species whose relationship with *Homo sapiens sapiens* would be extremely difficult or at least as difficult as ours are (or perhaps are not, given their extreme simplicity) with apes (Walker, 2011, 43). For his part, Max More states that transhumanism is both a reason-based philosophy and a cultural movement that affirms the possibility and desirability of fundamentally improving the human condition by means of science and technology. Transhumanists seek the continuation and acceleration of the evolution of intelligent life beyond its currently human form and human limitations by means of science and technology guided by life-promoting principles and values (More, 2011, 2013a).

Despite the great clarity of these statements, More had to provide further elucidation in order to avoid transhumanism being perceived as an unachievable utopia: Transhumanism is about continual improvement, not perfection or paradise. Transhumanism is about improving nature's mindless "design", not guaranteeing perfect technological solutions. Transhumanism is about morphological freedom, not mechanising the body. Transhumanism is about trying to shape fundamentally better futures, not predicting specific futures. Transhumanism is about critical rationalism, not omniscient reason. (More, 2011, 62). In these few sentences More firmly underscores his eminently transient vision of transhumanism. His principles focus on the means that enable constant enhancement, not on the state that may ultimately be achieved, and thus transhumanism can be recognised as having the same characteristics as those attributed to progress, which is nothing more than "a process of improvement that presents itself as linear, cumulative, continuous, necessary, irreversible and indefinite" (Taguieff, 2001, 19). James Hughes favours a more conscious and *dirigiste* interpretation of technological progress but he appears to share this opinion when he claims that "people are generally happier when they have more control over their own lives", although he identifies technology and democracy as the tools to achieve this greater control (Hughes, 2004, 9). The difference lies in the considerably broader aims that Hughes assigns to transhumanism, based on the ability to historicise the complex movement of Western societies towards greater freedom. Hughes clearly understands that there is no individual freedom abstracted from the historical conditions in which it is exercised and that, therefore, the promotion of political freedom brought about by a fair distribution of wealth and technology represents the underlying condition for mere individual freedom, that is to say of the possibility to achieve one's own life plan with all the means available and in all possible directions. The result is a shift towards sociological, political and economic themes absent in the first transhumanist declarations as these were entirely focused on the promotion and extension of individual freedom. Although these are significant differences, they do not impact the acceptance of a framework of practices, concepts and trends common to all transhumanism.

Nick Bostrom's definition of transhumanism is more specific. Although initially content to view the transhuman as a phase in the transition between human and posthuman, he offers an image of the human being of the future whose entirely transitory physical and biological features will be enormously different from those of today. As Bostrom clearly explains, transhumanism promotes an interdisciplinary approach to understanding and evaluating the opportunities for enhancing the human condition and the human organism opened up by the advancement of technology. Attention is given to both present technologies, like genetic engineering and information technology, and anticipated future ones, such as molecular nanotechnology and artificial intelligence (Bostrom, 2005c). The application of current and prospective technologies will impact on all stages of human existence: a longer and healthier life, improved rational faculties, refined emotions, a generally augmented individual wellbeing and, last but not least, the opportunity "to achieve a greater degree of control over our own lives" (Bostrom, 2005c, 87).

Wisely, Bostrom does not venture a definition of this final stage in which the posthuman will have reached its fulfilment, but he does explain that the transhuman will have at least eliminated old age and disease and improved intellectual, physical and psychological abilities. These new beings will not experience suffering, illness and ageing, they will enjoy a basically infinite life and they will have acquired an additional power that somehow collects and sustains all these changes: the freedom to give themselves any form they choose because they will enjoy what Sandberg defined as "morphological freedom" (Sandberg, 2003) through the use of technology. The problem with this view of transhumanism is the strong indication that the ultimate goal of enhancement is increased control over one's own life. Apart from the fact that there seems to be no guarantee that transhumanism can actually improve this specific aspect or indeed that this is its main purpose, it is the whole plexus around this notion of *control* that is questionable. Bostrom states that it is possible to significantly increase control over one's biology, and that wellbeing is directly proportional to this mastery of oneself. He assumes, however, that control and autonomy coincide completely and that the suffering of the human condition stems from a lack of control over our biology. Secondly, he overlooks the possibility that improved control of internal biological processes could have the unfortunate consequence not of abolishing uncertainty (who would not want the assurance that they would never grow sick or die?), but of introducing the human race into a predetermined future.

Transhumanists are globally and fundamentally technophiles and consider that technology is the main instrument of emancipation and liberation of human beings. Excluding its improper use by immoral or insufficiently skilled individuals, technology is a toolbox which can only produce positive effects, above all increased individual freedom. Technology is not specifically a political tool but is at the service of the advancement of individual freedom. The technologies promoted by transhumanism are generally identified by the acronym NBIC: nanotechnology, biotechnology, information technology and cognitive science (Roco & Bainbridge, 2002). Unlike other technologies humanity has used throughout its history to improve wellbeing and the performance of the different organs of the human body, transhumanists claim that NBIC technologies are more targeted and therefore faster and more effective, but, as we shall see, they also have many disadvantages.

References

Agar, N. (2010). *Humanity's End*. MIT Press.

Agar, N. (2014). *Truly human enhancement: A philosophical defense of limits*. MIT Press.

Bashford, A. (2013). Julian Huxley's transhumanism. In M. Turda (Ed.), *Crafting humans: From genesis to eugenics and beyond* (pp. 53–167). National Taiwan University Press.
Bernal, J. D. (1969). *The world, the flesh and the devil: An enquiry into the future of the three enemies of the rational soul* (1929). Indiana University Press.
Bostrom, N. (2005a). A history of transhumanist thought. *Journal of Evolution and Technology, 14*(1), 1–25.
Bostrom, N. (2005c). Transhumanist values. *Review of Contemporary Philosophy, 4*, 87–101.
Coenen, C. (2014). Transhumanism and its genesis: The shaping of human enhancement discourse by visions of the future. *Humana Mente: Journal of Philosophical Studies, 26*, 35–58.
Coeurnelle, D., & Roux, M. (2016). *Technoprog. Le transhumanisme au service du progrès social*. Fyp éditions.
Damour, F., & Doat, D. (2018). *Transhumanisme. Quel avenir pour l'humanité?* Le Cavalier Bleu.
Dard, O., & Moatti, A. (2016). Aux origines du mot 'transhumanisme.' *Futuribles, 413*, 154–165.
Deprez, S. (2019). Le transhumanisme peut-il échapper au libertarianisme? Le cas de James Hugues. *Raisons Politiques, 74*(2), 29–50.
Drexler, E. (1986). *Engines of creation*. Anchor Books.
Esfandiary FM (FM-2030). (1977a). *Up-wingers: A Futurist Manifesto* (1970). Popular Library.
Esfandiary FM (FM-2030). (1977b). *Telespheres*. Popular Library.
Esfandiary FM (FM-2030). (1978). *Optimism one* (1970). Popular Library.
Esfandiary FM (FM-2030). (1989). *Are you a transhuman? Monitoring and stimulating your personal rate of growth in a rapidly changing world*. Warner Books.
Goffi, J.-Y. (2015). Aux origines contemporaines du transhumanisme. In F. P. Adorno & J.-Y. Goffi (Eds.), Le Transhumanisme. *Ethique, politique, religions, 1*(6), 17–32.
Haldane, J. B. S. (1923). *Daedalus or science and the future: A paper read to the heretics cambridge*. Dutton & Company.
Hansell, G. R., & Grassie, W. J. (2011). *H+ transhumanism and its critics*. Metanexus Institute.
Harrison, P., & Wolyniak, J. (2015). The history of 'transhumanism.' *Notes and Queries, 62*(3), 465–467.
Hauskeller, M. (2016). *Mythologies of transhumanism*. Palgrave Macmillan.
Hottois, G. (2018). *Philosophie et ideologies trans/posthumanistes*. Vrin.

Hughes, J. (2004). *Citizen cyborg: Why democratic societies must respond to the redesigned human of the future*. Westview Press Books.

Huxley, J. (1957). Transhumanism. In *New bottles for new wine*. Chatto & Windus.

Kurzweil, R. (2005). *The singularity is near*. Penguin.

Lee, N. (2019a). *The transhumanism handbook*. Springer.

Lee, N. (2019b). Brave new world of transhumanism. In N. Lee (Ed.), *The transhumanism handbook* (pp. 3–48). Springer.

Marzocco, R. (2019). *Transhumanism. Engineering the human condition: History, philosophy and current status*. Springer.

More, M. (2010). The overhuman in the transhuman. *Journal of Evolution and Technology, 21*(1), 1–4.

More, M. (2011). True transhumanism: A reply to Don Ihde. In G. R. Hansell & W. J. Grassie (Eds.), *H+ transhumanism and its critics* (pp. 62–66). Metanexus Institute.

More, M. (2013a). The philosophy of transhumanism. In M. More & N. Vita-More (Eds.), *The transhumanist reader* (pp. 3–17). Wiley-Blackwell.

More, M., & Vita-More, N. (2013). *The transhumanist reader*. Wiley Blackwell.

O'Connell, M. (2017). *To be a machine: Adventures among cyborgs, utopians, hackers and the futurists*. Doubleday.

Roco, M. C., & Bainbridge, W. S. (2002). *Converging technologies for improving human performance: Nanotechnology, biotechnology, information technology and cognitive science*. NSF/Department of Commerce Arlington.

Sandberg, A. (n.d.). *We, borg: Speculations on hive minds as a posthuman state*. http://www.aleph.se/Trans/Global/Posthumanity/WeBorg.html. Last accessed 19 March 2020.

Sandberg, A. (2003). Morphological freedom: Why we not just want it, but need it. In M. More & N. Vita-More (Eds.), *The transhumanist reader* (pp. 56–64). Wiley Blackwell.

Sorgner, S. L. (2009). Nietzsche, the overhuman, and transhumanism. *Journal of Evolution and Technology, 20*(1), 29–42.

Stock, G. (1993). *Metaman: The merging of humans and machines into a global superorganism*. Transworld.

Taguieff, P.-A. (2001). *Du progres*. J'ai lu.

Tirosh-Samuelson, H. (2011a). Science and the betterment of humanity: Three British prophets of transhumanism. In H. Tirosh & K. L. Mossman (Eds.), *Building better humans?* (pp. 55–82). Peter Lang.

Vita-More, N. (2019a). History of transhumanism. In N. Lee (Ed.), *The transhumanism handbook* (pp. 49–61). Springer.

Walker, M. (2011). Ship of fools: Why transhumanism is the best bet to prevent the extinction of civilization. In G. R. Hansell & W. J. Grassie (Eds.), *H+ transhumanism and its critics* (pp. 43–50). Metanexus Institute.

Walker, M. (2014). *Transhumanism and Radical Enhancement*. http://eugenicsarchive.ca/discover/tree/545f9d2dbb4dbcc59800000. Last accessed 26 April 2020.

Young, S. (2006). *Designer evolution: A transhumanist manifesto*. Prometheus Books.

3

Transhumanism Between Humanism and the Posthuman

3.1 Introduction

Innate tendencies, long-enduring fears, refined biological theories and scientific and technological progress underpin the theories of the transhumanist imagination. However, the concepts contributing to transhumanism's underlying theories are much more complex than might be presumed. A vague reference to a desire for self-improvement, to fear of death or to the theory of evolution supporting transhumanism merely offers a descriptive acknowledgement of a theoretical movement that has had an impact of seemingly unprecedented magnitude and intensity. The role of philosophy and thought cannot be reduced to an objective description (although such a task might go well beyond simply recording information) but must necessarily perform the critical function of highlighting the implicit assumptions and the unstated consequences of this project, movement or whatever we choose to call it. In other words, a critique of transhumanism is needed in order to fully understand the underground lines of force constituting what is real and to render more complex and less evident those concepts and theories that at first sight seem natural, normal or clear.

F. P. Adorno, *The Transhumanist Movement*,
https://doi.org/10.1007/978-3-030-82423-5_3

Despite preaching hybridisation with artificial elements, transhumanism has the disconcerting property of attaching a certain degree of naturalness to its proposals. It is natural, they say, that humanity should desire to improve its physical and intellectual faculties because, after all, that is what it has always attempted to achieve; it is natural that it should seek the means with which to defeat death because that is what has always distressed it; it is natural that it should want to modify itself because that is what has characterised it since it first appeared on Earth. Transhumanism sets out to accomplish this and, therefore, merely reflects a natural tendency of humanity. It claims to offer nothing more than a realistic glimpse of humanity perfected and freed from all its unremitting anxieties and ills, something it has constantly longed for. In this sense, transhumanism is the theoretical continuation of the aims and objectives of humanity that have been debated and sought since time immemorial. In theorising enhancement through the bodily integration of exogenous devices or biological manipulation, transhumanism proposes nothing different from what the human race has always desired. Transhumanist enhancement is the development of a constant evolution of humanity that entails the introduction of artificial components more and more deeply within the human body. It is therefore unclear why theoretical justification should be needed for an activity that has always characterised nature even to the point of constituting the essential, specific element which differentiates the human race from all other animal species. However, all that is natural cannot, for that very reason, be historicised: to state that the transhumanist movement constitutes a movement that knows no discontinuity is, paradoxically, to conceal its origin. But is it fair to say that it is not a completely new movement, that there is no discontinuity in the history of humanity's desire to modify itself, that this tendency has no origin if not the advent of humanity on Earth? The moral and theoretical legitimacy of transhumanism clearly depends on answering these questions.

If it is true that enhancement/improvement has been the goal of all human actions since humanity's appearance on Earth, and is ultimately the driving force of science, technology and knowledge itself, then it seems unreasonable to criticise transhumanist supporters of enhancement when they claim that it is necessary and legitimate to carry on with the improvement of the physical conditions of existence, an undertaking that began explicitly with the advent of modernity and which can now truly improve people's lives through technological advances. While the opponents and critics of enhancement fear that humanity would fall headlong into a modern form of animality and nullify centuries of moral and material progress as well as the very "project" of Western thought, its supporters believe the opposite. They envision it as the tool for the advancement of humanity, a material improvement in life conditions whose ultimate goal is to heal Kierkegaard's sickness unto death and to create the preconditions on which to build a morally and materially better humanity as espoused by a movement that has grown in intensity over the last five centuries. To recognise oneself not only as a descendant of the philosophers of the Enlightenment and Renaissance humanism but also as one of those who seek to put into practice the trends and myths of humanity is to deny that transhumanism is "the world's most dangerous idea", as Fukuyama claimed, and be credited as a most fortunate continuer of hopes as ancient as mankind and perhaps one who will make them a reality.

Pursuing the transhumanist strategy of affirming the continuity of its proposals through the theories and practices of humanity ever since its birth presents two alternatives: criticising and denying all human culture along with transhumanism, or treating it in the same way as human artefacts whose moral legitimacy is never questioned. While the former would clearly be a difficult path to follow (and to what end, obtaining what results?), the latter would have extremely complex consequences, since the result would be an a priori justification of transhumanism de facto removed from any criticism. If humanity's nature were really to become a cyborg (anthropological continuity), how could transhumanism be criticised for proposing the cyborgisation of individuals? If transhumanism were the accomplishment of the ideals of humanism (theoretical continuity), the immediate consequence of discussing its aims would be a criticism of rationality and its eminently positive function.

These two alternatives are equally problematic, although for different reasons. If transhumanism is viewed as nothing more than the realisation of ideas and theories dating back at least as far as the Enlightenment and the birth of modern science, then any criticism of it would drag the whole rationalistic project of modernity into a destructive debate. On the other hand, if modern science and technology are considered as actually improving human existence by creating more comfortable conditions of life and reducing suffering and pain, then there seems to be no reason to eschew the prospects made available by transhumanism. However, a third, much more complex alternative would entail "saving" the past without yielding to the sirens of transhumanism, although this would require proof that the novel path laid out by transhumanism is nothing more than a deviation, a perversion or a pathology of the scientific spirit of modernity.

Identifying the "right" alternative will first require analysis of the relationships between all the theories whose denomination adds a prefix to the word "humanism". This concerns the relationships between humanism and theories that consider it incomplete or exhausted and hence strive to establish valid alternatives. The twentieth century was a period of reaction against and criticism of humanism: antihumanism, posthumanism,[1] ultrahumanism, metahumanism and transhumanism are some theoretical schools that expressed such criticism. But what is the relationship between them? To what extent do they converge or differ?

In order to avoid any confusion, one particular issue must first be dealt with. Transhumanism sees new technologies as driving the evolution of the human race, but there is no agreement—or at least no currently visible agreement—on the characteristic features of this evolution, that is whether it will lead to the creation of a new, posthuman species or whether this aim is desirable or inevitable and therefore passively endured. Of course, the most virulent criticism against humanism comes from the various complex positions of the theories of the posthuman, and transhumanism will have to justify its reasons for supporting or rejecting these. As transhumanism is a transitional stage along the path towards posthumanity (whatever its characteristics might be), it will be important

[1] From now on, by the term posthuman I mean the theories of the posthuman, an expression that will sometimes be used in its entirety for a better understanding of the text. The same term in italics without further specification, *posthuman*, I will use to identify an anthropological category: the human being of the future, however we may want to understand it.

to determine whether transhumanism falls within the general *posthuman* category.

Second, does transhumanism establish a relationship with humanism and, if so, how? According to its supporters, transhumanism is merely the continuation of humanism or, as Cary Wolfe acknowledges, "an *intensification* of humanism" (Wolfe, 2010, xv). However, greater care will be needed in analysing these assertions: what does "transhumanism is an intensification of humanism" actually mean? What exactly is an intensification? Does this refer to certain elements of humanism (and, if so, which ones?) or to humanism as a whole?

Transhumanism certainly fits into the category of theories that have a more or less dialectical relationship with humanism, as its founders have acknowledged and as is clear from the name itself. The twentieth century was a period in which the value of humanism was discussed, criticised and relinquished around the world, but also taken up again and defended with a tenacity at least equal to its value over the centuries. What position does transhumanism occupy in this multifaceted, theoretical panorama?

The first step in answering this question will entail defining humanism as comprehensively as possible. Then the theories of the posthuman will be thoroughly investigated so as to identify its essential characteristics, starting with its typical criticisms of humanism. Third, explanations will be given as to why transhumanism does not fall within the posthuman framework but instead represents ultrahumanism. Finally, transhumanism will be compared with humanism through an analysis of some key concepts and, in particular, notions from the semantic field of progress.

3.2 Humanism

The history of humanism is extremely complex, and it is certainly impossible to summarise it here. However, some general considerations can be provided in order to glean the elements present in most of the definitions and discussions that have taken place so far. Since Aulus Gellius, humanists have been regarded as people dedicated to a certain type of study, having been schooled in the liberal arts (the *trivium*, comprising rhetoric, grammar and dialectics, and the *quadrivium*, comprising music, geometry, arithmetic and astronomy), and therefore devoted to the promotion of the values of *humanitas*. In other words, according to the ideal of stoic Roman humanism, education aims to create a man endowed with virtue, rigor, *pietas*, austerity, amiability, *gravitas*, *severitas*, *dignitas*, *facilitas* and *clementia*, which distinguish him not only as a man of value but also as a moral and political model and thus worthy to exercise a number of functions within the *societas*. Education ennobles a man by teaching him moral and intellectual rectitude, which are manifested in the urbanity of his manners. *Humanitas* is an excellent way of being human, but it is also the way that leads to *humaniter vivere*, a condition that detaches man from a savage lifestyle in the state of nature.

It is this conception of education and training that was revived in the eighteenth-century Enlightenment and that nineteenth-century historians would identify in the Renaissance, when classical values and culture were rediscovered (Davies, 1997, 9–10). The German *Bildung*, inspired initially by the ideals of classical humanism and subsequently by Renaissance humanism, highlights the need to promote humanity in every single member of society in order to make it more humane. This form of humanism based on the rediscovery of classical Greek and Roman culture and practised in fifteenth and sixteenth-century Europe (especially Italy) was certainly an attempt at forward-looking cultural renewal achieved by learning from the past. But more than this it conveyed a particular conception of humanity and of the individual in possession of a number of qualities acquired throughout life and structured around such concepts as freedom, autonomy and independence, which were to be taught and promoted in order to improve the everyday life of the individual and of society as a whole. According to the latter conception, humanism emphasises the value of human rationality, promotes intellectual autonomy and individual freedom, protects moral independence from prejudice of any kind, and recognises the value of progress in all its forms (Davies, 1997, 22).

Condensing these two conceptions, it can be said that Renaissance humanists were inspired by ancient Greek and Roman societies in which the use of reason underpinned the struggle against ignorance and superstition (Davies, 1997, 22–23). The humanists fundamentally sought to improve humanity, or rather to improve the spiritual, moral and political dimensions that would make society more agreeable, simpler, less brutish and more "humane". However, humanism was not limited to the sphere of intellectual improvement but had a wider impact: the same scientific progress it promoted highlighted the need not only to strengthen human psychology but also to analyse, study and improve the more properly zoological aspect. In short, education in humanity was undeniably fundamental in tempering the brutishness of the human animal; but because humanity is also—and above all—an animal, the biological improvement of humanity could not be neglected. Humanism also gradually acquired the awareness that the teaching of virtues was fundamental but also that eliminating pain and illness was equally important and that moral progress could increase exponentially thanks in part to material progress.

However, these references are certainly insufficient in identifying the myriad theoretical threads that have intertwined to give life to the contemporary concept of humanism, underpinned as it is by an autonomous and independent conception of humanity and its ability to judge, behave and live. Indeed, nineteenth-century historians who elaborated on this image claim that this ability is particularly evident in Giovanni Pico della Mirandola, regarded by transhumanists as one of their predecessors. In the opening pages of *Oratio* Pico explains how it is impossible to give a positive description of humanity: its nature is not to have a definite nature, but to be able to choose what it wants to be. Humanity occupies a unique place in the animal kingdom because it is the only creature able to decide what it wants to make of itself and its life, given the plasticity and indeterminateness that distinguish it (Cassirer, 1948, 224).

This discretion may be a source of improvement but also of degeneration: if humanity has the freedom to elevate itself to the level of angels, it may also run the risk of lowering itself to that of other animals. Clearly, this attribution is a double-edged blade: with great freedom comes great responsibility for what one is. But regardless of the choices humanity makes as an individual and because it enjoys such great freedom accompanied by the faculty to choose what is best for it, it rises above all other animals and thus deserves respect, protection, praise and perhaps even worship in a religion centred on humanity. In short, humanism elevates humanity to a universal model, a benchmark, and by specifying what a human being is it simultaneously provides a normative criterion for defining dignity and specifying who should be respected, and who or what should be defended and valued. The image of the Vitruvian Man, which has become a representation of humanity as an image and model of nature, is not only the emblem of a precise historical period and a theoretical movement, but it is also the ductile representation of a specific conception of rationality and its morally normative force (Braidotti, 2013, 13).

This will hopefully have provided the reader with an albeit concise overview of humanism that practically all theorists of the overcoming, exhaustion and/or fulfilment of humanism would share. Anthropocentrism and universalism characterise a person who has been brought up on the Latin texts, has been taught to use rational thought[2] and is free to give themselves the form they want. However, this model of humanity weighing up all things and providing a benchmark for right and wrong is two-sided. On the one hand, humanity exhibits its very best characteristics but, on the other, in claiming to be representative of humanity, it exerts violence on others and casts into a brutish limbo all those humans and other creatures guilty only of being different and not sufficiently "human".

3.3 The Theories of the Posthuman

This consolidated theoretical basis fuelled the increasingly acknowledged realisation that the humanist dynamic had waned. It was now necessary to find new alternatives, to portray humanity in a different light and to

[2] A further explanation can be found in a famous quotation by Kant: "The fact that the human being can have the 'I' in his representations raises him infinitely above all other living beings on Earth. Because of this he is a person, and by virtue of the unity of consciousness through all changes that happen to him, one and the same person—i.e., through rank and dignity an entirely different being from things, such as irrational animals, with which one can do as one likes" (Kant, 2006, 15).

generate a conception of it that was less ideological and better at encapsulating a multiplicity of dimensions while also relinquishing the normative and normalising intentions of humanism. Renaissance humanism was the bearer of positive values (autonomy, independence, critical capacity, rationality and so on) but also negative ones (the ideology of rationality, racism, sexism, selfishness, the exploitation of nature and so on) and, as the latter were preponderant, the resulting damage was so blatant that a rigorous critique of humanism began to emerge in an attempt to identify different and less pernicious ways to describe the human being. These critiques resulted in different forms of antihumanism during the twentieth century, ranging from those of philosophers such as Althusser or Foucault (Soper, 1986) to those promoted by feminist movements, technological studies (Haraway, 1991) and cultural and race studies, which were in turn subjected to more or less justified virulent criticism.

The main criticism lodged against humanism is its vision of the human being as a being who autonomously places itself at the centre of the universe to act as a benchmark for all things. The fact that it is perfectly rational, or at least capable of a coherent and continuous use of its rationality, makes it autonomous and independent. Antihumanists have mainly decried the practical consequences of this image of the human being because it has led to forms of discrimination against those who do not match up: women, black people, slaves, disabled people, animals and so forth, who are considered as deviants. In other words, they point the spotlight at the imperialism and the cultural and theoretical colonialism of humanism that led to forms of exclusion, exploitation and subjugation of others (Davies, 1997, 23).

Antihumanism in its various forms has objected to anthropocentrism and the way in which the human being was defined according to particular characteristics. Instead they proposed an alternative image of the human being, a person who has developed beyond the canons of Western cultural tradition: a *posthuman*, a term that entered the theoretical and philosophical lexicon in the late twentieth century (Hassan, 1977). However, posthuman is merely an umbrella term that brings together a series of somewhat miscellaneous theories and views predicated on the awareness that humanism is now obsolescent and that the view of the human being as a benchmark for all things or even the dualism of humanism cannot adequately describe human reality (Ranisch & Sorgner, 2014, 16). The posthuman is a critical movement that targets the hegemonic image of the individual (the result of combining Cartesian dualism with Locke's possessive individualism) employed to corroborate the control practices and education (or child-rearing) policies in Western liberal societies. The features that all forms of the posthuman share are their criticism of the humanist conception of the human being and their intention to deconstruct its subject and demonstrate that the alleged centrality of humanity is an illusion.

This seems to confirm the view that the posthuman is, at first sight, fundamentally connoted by the desire to overthrow human sovereignty by exposing it as a contrivance and revealing how it conceals a radical and inevitable hybridisation with animal life forms and artificial mechanisms (Nayar, 2013, 11). Such changes or modifications of the human being have indeed been proposed, as can be seen in the term *posthuman* used in the catalogue of an exhibition in Lausanne in 1992. The gallery owner and curator Jeffrey Deitch wanted to define a general trend towards the creation of new characteristics structuring subjectivity, and thus the search for new forms of personal identity. Art models its objects along the same lines that lead to the construction of the subject, and is perhaps applicable to the subjects themselves, which thus become artistic objects. Indeed, Deitch claims that a parallel can be drawn between "new approaches to self-fulfilment" and "new approaches to art". Artists would thus be catalysts for social change, which would in turn impact on the creation and definition of new personalities developed outside the assimilation of hegemonic social, technological and political models (Deitch, 1992). This view of the *posthuman* as a unique opportunity to reshape the human physical form, partly by reviving archaic practices of body modification, quickly combined with a form of technophilia and has now been put into practice, sometimes exaggeratedly, by such well-known artists as Stelarc or Orlan. This should come as no surprise because the *posthuman* does not only connote other models of subjectivity and does not consider humanity merely as part of an ecosystem within which it evolves along with other life forms, but as a being that incorporates artificial mechanisms and technology into its biological substance (Nayar, 2013, 19). The theory of the posthuman thus aims to overcome humanism because it does not consider the human being as a collection of characteristics such as rationality, authority, autonomy and agency exclusive to a universal subject, nor as the holder of certain values conferring a dignity that sets it apart in the animal kingdom, nor as a single entity in the natural and biological landscape, but as an organic assembly that evolves with other vital forms (Nayar, 2013, 14).

Posthuman theorists argue that the image of humanity proposed by humanism not only never existed but is actually useless, rendered obsolete by events and by technology enabling new forms of existence. Consequently, reason must likewise reject the categories adopted by humanism or during the Enlightenment and produce novel ways of accounting for this completely different situation. It is also argued that the human body is already a hybrid, a mixture of elements taken from animals or developed by technology, and it is thus not natural and has always been a technological body. In short, the posthuman is a set of theories that probes new forms of existence, analysing their characteristics and testing their consistency and viability from a practical, ethical and theoretical point of view, starting from the assumption that, ontologically, humanity is a hybrid being and that this aspect has always been concealed. For this reason, the posthuman is *also* an essentially critical movement (Nayar, 2013, 12).

A definition of the person proposed by the critical or philosophical posthuman must necessarily start from its primary characteristic, which is undeniably the morphological freedom achieved through forms of hybridisation with the surrounding environment. A second characteristic is a non-qualitative and normatively neutral conception of difference: even if there are clearly differences from a biological point of view, the theory of the posthuman does not differentiate between human, inhuman, non-human and/or animal beings by placing them on a scale that measures their qualities or the extent to which they resemble an anthropological ideal. Third, this person has the willing to cooperate and hybridise with other forms of life or simply with the natural and artificial elements of the surrounding environment.

3.4 Is Transhumanism a Form of the Posthuman?

As already mentioned, "posthuman" is an umbrella term encapsulating different views and theories. Critical and cultural posthumanism is flanked by technological posthumanism (Tirosh-Samuelson, 2014), and it is precisely this overlapping of beliefs and convergence on certain forms of technophilia that result in its being confused with transhumanism. For both movements it is a question of the posthuman human. For critical or, in some way, postmodernist posthumanists, it is a question of breaking away from the ideals of classical humanism by defining a non-hierarchical, widespread, hybridising and hybridised form of humanity. But is transhumanism heading towards this form of the posthuman? Does technophilia have the same purpose and produce the same results in both cases? If we accept the prefixes, the only answer is "yes": the transitioning person of transhumanism must be heading towards the *posthuman*, that is what comes after humanity. However, posthumanists do not think at all about the *posthuman* implicitly sketched out in transhumanist texts but instead see the *posthuman* of transhumanism as the fulfilment of all the ideals of Western humanism. Nevertheless, there is a possible source of ambiguity in the convergence of the one and the other on the definition of a hybrid human being. Transhumanism promotes an anthropological ideal very distant from that of the critical (and cultural) theories of the posthuman, although they do have at least two elements in common: morphological freedom and hybridisation with a variety of elements extraneous to human biology, which will be discussed below. Will this suffice to claim that there is a convergence of views between the one and the other? Absolutely not.

The difference between a cultural and theoretical, deconstructive and critical theory of the posthuman and a technological and assertive, empowering and practical theory of the posthuman, a sort of ersatz transhumanism, was suggested by Katherine Hayles. In her view, anti-humanist posthuman theory can be defined with reference to four contributing factors. First of all, the posthuman fosters a conception of humanity as an information hub rather than as a fundamentally biological being, something that is held to be completely incidental. Second, it follows that consciousness is not the primary constituent element of human identity, as Descartes claimed, but an epiphenomenon, an entirely secondary product of evolution. Third, the body is nothing more than a prosthesis comprising totally interchangeable modules in a process which began at birth and will only be completed when the body dies. Finally, and most importantly, the body spreads throughout the world: the boundaries that seem to isolate it from the environment in which it lives are cultural contrivances that do not actually exist. We are not *in* the world—we *are* the world (Hayles, 1999, 2–3). Seemingly contradictory ideas are present within these theories. The posthuman certainly makes it possible to set aside the idea that the subject is natural, but the constructivism of subjectivity does not rule out the possibility that the final result could be a cyborg like the Six Million Dollar Man or RoboCop, who seem to represent the ideal of the transhumanism.

Hayles, on the other hand, stresses the importance of not seeing the cyborg as the only and, ultimately, the most valid representation of the *posthuman.* While it is true that posthuman theory suggests the emergence of new models of existence and subjectivity that incorporate artificial modules within them (Hayles, 1999, 3–4), this is regarded as a possibility and not a necessity. While Katherine Hayles hopes that the posthuman movement is above all critical of liberalism and individualism, she believes that the emphasis placed on the importance of technologies in posthuman theory may lead to some confusion, underlining the need to make a clear distinction between the posthuman and transhumanism: If my nightmare is a culture inhabited by posthumans who regard their bodies as fashion accessories rather than the ground of being, my dream is a version of the posthuman that embraces the possibilities of information technologies without being seduced by fantasies of unlimited power and disembodied immortality, that recognises and celebrates finitude as a condition of human being, and that understands human life is embedded in a material world of great complexity, one on which we depend for our continued survival. (Hayles, 1999, 5; see Dupuy, 2011) Given these premises, it is true that the posthuman could have two consequences, but there is nevertheless a fundamental distinction between the posthuman and transhumanism: the former is a disempowering movement that acknowledges, understands and accepts the complexity of the world in which human beings evolve, while the latter is an empowering movement that ignores everything that does not serve its own affirmation.

If we accept the view expressed by Hayles, a leading exponent of critical and cultural posthumanism and attentive to all aspects of theory, it appears incontrovertible that transhumanism and the posthuman are opposed on one absolutely fundamental aspect even though they may share others. But what is the position of transhumanists regarding posthuman theorists like Hayles or feminists like Donna Haraway with whom they might be confused?

Bostrom sketches the possible characteristics of the posthuman developing along transhumanist lines. If we adhere to the view that transhumanism fundamentally supports human empowerment through hybridisation with technology, we can also see a broad outline of this possible person of the future. However, we cannot do much more because we lack the cognitive abilities to achieve a complete understanding of this new human being, just as apes lack the ability to understand the specific nature of humanity (Bostrom, 2005c). But we know that it is the enhancement of humanity, that is a person enhanced by technologies that will infinitely upgrade its performance and conditions of life. It will have almost total control over its biological processes and will therefore be able to dominate and modify them, free from pain, illness, suffering, death and the limitations imposed by its constitution: it will indeed be the kind of person who haunts Hayles's dreams, turning them into nightmares.

Transhumanism can therefore be collocated in the panorama of theories that engage with humanism. It does not entail forms of transience typical of a cultural movement centred on the human being and its characteristics but the potential modification of a biological entity, which is what the human being is; it is not a cultural movement that analyses the specific nature of the human being but a movement that is concerned with highlighting the ways in which technology or science concretely impact the human being as a biological entity. Another useful criterion for defining the specificity of transhumanism is that the indicated direction of change is completely different from those of the movements discussed above. There are certainly some issues in this regard because this change concerns the very nature of the human being, of which we have an extremely vague and imprecise idea, to use a pale euphemism, and which should be transformed into something more than human, a *posthuman* with which we are totally unfamiliar even though we may envisage a living species that, by definition, cannot be stabilised.

Regardless of any similarities, therefore, there is one fundamental aspect setting transhumanism apart from the critical, theoretical or philosophical posthumanism: its declared intention to modify the human being. The posthuman has a critical, aesthetic and, perhaps, ludic attitude, creating an anthropological figure inspired by Aristotelian praxis. The exercise of human freedom lies in the creative process, not in the final product. Transhumanism has a progressive, positive attitude and tends to create an anthropological figure inspired by Aristotelian *poiesis*: the product is a symptom of human freedom, not the production process. The freedom invoked by transhumanists is the consequence of the liberation from a biological condition that distinguishes the human race as living beings. The fundamentally innate alienation of the human being can be absorbed through access to forms of biological invulnerability, that is incorporeality or, better still, immortality. Thus, according to the transhumanist view, the term "posthuman" does not denote a break with the cultural movement of humanism but signals the contestation and rebellion against the inevitability of the constraints of human biology, in line with a movement that had already been established but which only gained momentum in the twentieth century thanks to advances in nanotechnology, biotechnology, information technology and cognitive science, particularly biotechnology.

3.5 Is Transhumanism a Form of Humanism?

At this point it is reasonable to inquire about the connection between transhumanism and classical and Enlightenment humanism. This is an important question because, as stated previously, part of transhumanism's claim to theoretical legitimacy relies on its filiation: viewing transhumanism as a modern invention would have this immediate consequence. Defining it as a branch of movements and tendencies consubstantial to the human being, that is making it a somehow human specificity, would debase or utterly eradicate any demand for a justification of its theories. Clarification is therefore needed on the individual envisioned by transhumanism: is it a continuation of the human being of humanism (Hauskeller, 2016, 97–129), an intensification of it, a pathological betrayal of it or a completely innovative figure breaking with all that has gone before?

At first sight, transhumanism would seem to stem from the same awareness and desire that drove the modern scientific revolution. If this initial impression were confirmed, there would be no reason to contest the transhumanist claim that they belong to a theoretical movement promoting and justifying the improvement of human nature that was developed during Renaissance humanism by thinkers like Pico della Mirandola and which has continued up to almost contemporary philosophers. However, in order to determine whether there has been a break with humanism it will be necessary to scrutinise the various details, concepts and theories one by one. Tackling these issues with reference to broad categories such as "humanism" is of little help, which is why this study aims to analyse humanism through its individual constituent elements. It is in relation to these elements that we can identify any instances of continuity or seek and define any differences.

Transhumanists (Bostrom first and foremost) accept humanism's promotion of rationality as well as its desire to increase freedom, democracy and tolerance, and they consider respect for human beings as a value in itself. However, they argue that all this can be achieved through less classical means than those proposed by humanism (education and culture) and specifically through the far swifter and more effective application of biotechnology. Transhumanism is thus a stage on a path aiming to implement the humanist project, the means and scope of which (health, cognition, emotions and life expectancy) have been revived and enhanced. This willingness to progress constitutes the fulfilment of the human beings as a biological entity but also of humanism as a cultural movement centred on the promotion and refinement of the specific characteristics of humanity. Far from being a source of discontinuity and an interruption of a becoming that branches into a different and not yet theorized future, transhumanism sees itself as the completion of a theoretical path originating in Renaissance humanism, some fundamental themes of which it explicitly embraces, such as human freedom understood as the "formal" cause of self. It remains to be seen whether this self-definition corresponds to the truth, but this does not detract from the fact that transhumanism (and the posthuman which ordains its completion) might perhaps constitute yet another variant of Renaissance utopias.

Secondly, from the point of view of anthropology proper, the transhumanist ideal seems to be in clear continuity with humanist ideology. The individual of classical doctrine is autonomous, independent, rational and free, as is the one that transhumanism aims to create. The substantive difference lies in the means by which this goal is to be achieved: biotechnology instead of reading, genetic manipulation instead of education in thinking, hybridisation with artefacts instead of exercise and practice. Is this difference sufficient to brand transhumanism as extraneous to the cultural, theoretical and scientific movement of modernity? The answer is necessarily complex and comprehensive, not least because it has to take into account the role of technology and its moral impact on the development of the human being. For the time being it will suffice to say that the purpose and the features that the different modes of human development must shape are the same. Paradoxical as it may seem, there appears to be a certain continuity on this point with the Enlightenment. As Bacon inquires, should man not make himself the lord and master of nature through knowledge of its processes? Should he not extend his dominion over nature through his increasingly refined knowledge? Is the human body and the whole human being not part of nature and, like the rest of nature, should he not be better known for the purposes of his dominion and control? This is what Spinoza also assumes when he warns against considering "man in nature as a dominion within a dominion" (Spinoza, 1985, 491). If human actions are subject to the same laws that govern the rest of nature, it seems unthinkable that the dominion over nature sought by all the means of modern science should refrain from exploring human processes so as to then dominate them as it does for all things that do not derive from human action. Bacon himself, in a list entitled "*Magnalia naturae praecipue quoad usus humanos*" clearly sees in science a number of objectives that could easily be considered to be those of transhumanism.

A problem arises at this point. Should transhumanist enhancement aim to augment or improve human faculties to their physiological limits, as in biomedical interventions, or should it push the enhancement of existing human faculties beyond their biological limits, thereby creating a different and above all superior species? As far as the relationship with humanism is concerned, it would be one thing to state that the purpose of human enhancement is to improve human faculties and functions up to the limits of their biological possibilities—an objective compatible with humanist anthropology—but it would be quite another to claim that enhancement should seek to create new capabilities, as this type of enhancement would or could lead to the creation of a species of being that is better than the one currently defined as human: in other words, a veritable posthumanity. It is therefore absolutely essential to establish whether transhumanists favour "human" enhancements, that is ones that, at first sight, do not create a new species but only an augmented humanity, or "superhuman" enhancements through which a new species would be created, that is *posthumans* that would resemble *Homo sapiens* to the same extent that *Homo sapiens* resembles apes. In short, the transhumanism proposes a moderate enhancement, which takes a number of functions to their current limit but without exceeding it or a radical enhancement which takes these functions to an unknown and currently unreachable level (Agar, 2014, 2; Mehlman, 2000, 523)? Understanding which type of enhancement transhumanists advocate has a major impact on an argument they bring up to legitimise their aims. If radical enhancement is promoted, it would indeed be difficult to justify transhumanism by claiming it is carrying forward the attempts that humanity has made to enhance its condition since the dawn of time, and in particular from the seventeenth century onwards. However, in the absence of this continuity, transhumanism would lose one of the most powerful arguments in favour of its legitimacy. Furthermore, differentiating the implicit or explicit aims of "transhumanist" enhancement from those proposed by "humanist" enhancement is an absolutely key step because criticism of the transhumanist project sometimes targets the idea that enhancement is a practice aimed at improvement, empowerment and the augmentation of already existing functions and organs, while at other times the focus is on the future characteristics of a *posthuman* of which only a very blurred outline is visible for now.

Unfortunately, the transhumanists themselves are divided on this point. On the one hand, Max More is extremely clear: transhumanism goes well beyond humanism in both means and ends. [...] By thoughtfully, carefully, and yet boldly applying technology to ourselves, we can become something no longer accurately described as human—we can become posthuman. Becoming posthuman means exceeding the limitations that define the less desirable aspects of the "human condition". (More, 2013a, 4) Even Bostrom is not always clear about the aims of enhancement; he wavers between the idea that it should be limited to upgrading human faculties beyond the generally acknowledged and statistically measured standard, for example raising the IQ beyond the maximum, and the alternative view that it should be extended to the addition of faculties, organs and functions that are not human, such as wings or sonar. Bostrom not only seems singularly uninterested in probing this difference, but he even seems to focus on "humanist" forms of enhancement which produce an improvement in currently existing individuals (Bostrom, 2009a).

Clearly, these two occurrences will not suffice to provide a clear and precise idea of the transhumanist position on a key aspect, on which not only the characterisation of transhumanism but also its moral and theoretical legitimacy depend. However, there seems to be no simple solution, not only because of specific shortcomings on the part of transhumanists but because of an inherent difficulty. For example, a *posthuman* who underwent a radical mood enhancement would be so neurologically different from other people that the latter would be virtually unable to understand the emotional states expressed by this being. In other words, they would find themselves in the same situation as young children trying to understand adult emotions that they have never experienced and of which they do not and cannot have any knowledge (Bostrom, 2009a, 11). However, if it is true, as Bostrom points out, that one cannot criticise what one does not know, then it is equally true that advocating what one does not know is senseless. As mood enhancement is utterly beyond human perception and knowledge, it would be advisable to avoid espousing its characteristics, even generically (Hauskeller, 2013, 21). In other words, if transhumanist enhancement were radical in nature, it would no longer be subjected to the criticisms made of humanist enhancement. On the other hand, an over-decisive focus on the difference between moderate humanist enhancement and radical hyperhumanist enhancement would render the transhumanist project incomprehensible. While we await the arrival of a future human race about which we know nothing except that it will be different and better than today's, interest in transhumanism and the very meaning of enhancement are dependent on what human beings are able to accomplish here and now, and they use this basis to sketch out an albeit imprecise picture of the direction in which the future of humanity is travelling. For this reason, we will somehow be forced to overlook the difference between transhumanist and humanist enhancement, or at least to consider transhumanism as hyperhumanism, since only of the latter is it possible to form an idea and develop an analysis.

Thirdly, and from the point of view of the philosophy of history, there seems to persist a certain continuity between eighteenth-century and Enlightenment humanism and transhumanism. It does not seem unreasonable, therefore, when the transhumanists claim that they embrace the Enlightenment project of Condorcet, for example, who has been cited on numerous occasions as a precursor of their theories. In *Sketch for a Historical Picture of the Progress of the Human Mind* and in a fragment on *Atlantis* partially integrated into the 10th Epoch of the Sketch, Condorcet assigns the same goals to science as the transhumanists, which match those already identified by Bacon: to find a way to extend human life, strengthen human faculties, eliminate disease and control population growth. The enhancement process aimed at improving human virtues will be even more effectual because it will change human nature by making it better disposed to do good. The rootedness of perfectibility in the physical organisation of the individual is an implicit assumption of transhumanism and the first step in this new scrutiny of the relationship between the physical and the moral—one of the topoi of nineteenth-century French philosophy and, in particular, the *idéologues* and the so-called *médecins-philosophes*. Starting from this desire to link the physical and the moral, a direct consequence of the criticism and the attempt to dispense with Cartesian dualism, Condorcet can legitimately envision an indefinite enhancement in the material conditions of life underpinning the moral enhancement of the individual. Since the relationship between physical organisation and virtue seem real, the physical enhancement of humanity must necessarily lead to the constitution of a more virtuous human being. In short, for the Enlightenment, or at least for certain Enlightenment thinkers, the liberation of the human being necessarily passes through the alteration of his body, that is through the manipulation of the living person. As if these extreme similarities between Condorcet's theories and the aims of transhumanism were not enough, it must be underlined that one of the fundamental elements of transhumanism is its focus on technological and scientific progress and thus a certain view of history and of becoming, which is formulated for the first time in this period.

An analysis of the concept of progress (and the normally associated or assimilated concepts of perfecting, perfectibility and perfection) cannot be adequately performed here. However, it should start from Bacon and the subsequent *querelle* between the ancients and the moderns in which the quest for the best, for the "more than good" or, to use a term that begins to acquire a temporal meaning in this period, for progress, is more than evident. Obliged to explain why the present is better than the past, the supporters of the modern are inclined to devise a philosophy of history which, while not exactly based on a teleology, must at least propose a specific notion of progress. From this moment on, progress begins slowly but surely to lose its moral connotation and its relationship with the idea of perfection, with which it has long been synonymous. For Greek and Roman philosophy, or even for Christian theology, becoming perfect, or at least committing oneself to this path, meant voluntarily reaching a clearly defined state, progressing to a certain point where every moral objective had been fully achieved. Progress was viewed as a moral intensification of individual conduct aiming at the attainment of a final state of perfect knowledge and an equally perfect love of God, rather than a progressive material improvement. The process of moral perfection is therefore completely different from historical progress. On this individual path towards perfection, there is no "new" but simply an intensification, a deepening and an expansion of what already exists. The effects of this conception of progress are particularly evident in Condorcet. His introduction to *Sketch*, where the purpose of knowledge is identified as the capacity to accelerate the progress of humanity, forcefully sets out the clearly Baconian theory that knowledge has an operative purpose because it is a tool for use in the modification of nature. The observations that Condorcet seeks to develop in the *Sketch* "on what man has been and what he is today" must aim to "ensure and speed up those new advances that his nature still makes it possible to hope for" (Condorcet, 1955, 125). The first result that Condorcet aims to achieve is to show that "perfectibility is truly indefinite" (Condorcet, 1955, 176), that it has no limits assignable a priori, that it cannot be stopped, that it cannot be changed. But it is at the end of his *Sketch* that Condorcet better clarifies the sense of indefinite perfectibility, where its already weak and almost completely devitalised moral connotation is rapidly absorbed by scientific necessity.

The reduction of perfection to progress is thus definitively accomplished. Progress was synonymous with perfection prior to the sixteenth and seventeenth centuries but they became two radically different notions in the seventeenth and eighteenth centuries and then, in the eighteenth and nineteenth centuries, perfection became a synonym of progress. But this has come at a price: perfection must be uprooted from its theoretical soil (comprised of a certain conception of ethics and morality) in order to be grafted onto the idea of indefinite, continuous and necessary progress that lay at the origin of the competing concept. In other words, perfection is reduced to progress through the elimination of a purpose, which is one of the specific characteristics constituting modernity. However, although the notion of perfection will become increasingly problematic in modernity, when its derivatives will acquire increasingly complex meanings, progress and perfection have almost come to resemble two dry leaves on a now dead tree. This is the cultural and theoretical phenomenon encountered when we assign a meaning to the contemporary term "progress" and especially to that which appears to be implicitly operative in the philosophy of the history of transhumanism. Progress has no spiritual or moral dimension for transhumanism but is simply a process with no goal or purpose other than the "indefinite" improvement of individual performance. Moreover, in accordance with its modern origin—which entwined the physical and the moral, biology and history, anthropology and humanism—transhumanism maintains that the driving force behind any progress can only be an enhancement of the biological basis of humanity, of its "physical" aspect, insofar as this is an essential condition of all "moral" processes.

This overview of some concepts common to both technological-scientific rationalism and transhumanism can now be followed up with some conclusions. First, there is a sort of diagnosis common to the two movements regarding the physiological and moral state of humanity that may be on the verge of degenerating. It is said that if we do not wish to decline, we absolutely must make an additional effort to improve and enhance our faculties. The reasons for this decline are identified in various phenomena: some claim these lie in the advent and spread of the industrial society, while others suggest it is because natural resources have been overexploited. There remains a shared awareness that humanity is under threat, if not of extinction then certainly of disempowerment. Second, this diagnosis claims that humanity needs biological improvement in order to enhance its performance and adapt it to the challenges that await it. The means of achieving this are inevitably different due to the different technological and scientific progress in these two periods; but Bacon, although unaware of the means needed to reach this goal, saw science and technology as fundamental because these aims could only be achieved through biological manipulation. Third, history is subject to the same dogged determinism as nature: Auguste Comte, Max Moore and Ray Kurzweil seem to agree on the relationship between freedom and history. Historical laws are analogous to natural laws, and historical becoming can be read essentially according to technological and scientific progress. For Comte the human being in its individuality and humanity in its generality go through three successive stages (theological, metaphysical and positive) according to their degree of perfection: the link between past, present and future in a continuum culminating in the achievement of a post-political stage governed by scientists and "industrial works managers" is determined (Comte, 1996, 266). Because of their knowledge, the latter will be the only people suited to govern a society geared exclusively towards production, and able to modify nature for the benefit of man. Natasha Vita-More recently recalled the importance for transhumanists of Gordon's law, which establishes and predicts that the processing power of computers will double every two years (Vita-More, 2019a, 52). But the most obvious analogy is between Comte and Condorcet on the one hand and Kurzweil on the other. Like Condorcet, Kurzweil describes an exponential progression of knowledge and technology in six stages, the last of which will be the achievement of *singularity*, marking the end of human evolution (Kurzweil, 2005, 38). As in Condorcet, this theory can be proved inductively from observation of human history. As in Condorcet and especially in Comte, Kurzweil also sees history as following laws that doggedly determine its becoming. Attempts to deviate from them, ignore them or deliberately break them could lead to the degeneration or disappearance of mankind. Enlightenment thinkers and transhumanists agree that humanity has all it needs to improve human existence, but this can only be accomplished if technologists and scientists are in charge.

In the final analysis, the continuity between transhumanism and humanism does not seem to be in question: on the contrary, the former stands as a hyperhumanism, an intensification of the ideals and tendencies of humanism. We will see if the continuation of this analysis will confirm this initial conclusion. If this is the case, it will have a significant consequence from two points of view: on the one hand, it could be argued that any critique of transhumanism should involve humanism or at least some undeniably important concepts of humanism; on the other hand, criticism of transhumanism would cast a rather sinister light on humanism which, unless historically contextualised, would lose any emancipatory force and would be reduced to nothing more than a movement that confused the moral and political improvement of humanity with material progress, entwining them in a deadly embrace that transhumanists have exploited to promote an individualistic ideology of performance and consumption.

References

Agar, N. (2014). *Truly human enhancement: A philosophical defense of limits.* MIT Press.

Bostrom, N. (2005c). Transhumanist values. *Review of Contemporary Philosophy, 4*, 87–101.

Bostrom, N. (2009a). Why I want to be a posthuman when I grow up. In A. M. Cutter, B. Gordijn, G. E. Marchant, A. Pompidou, & R. Chadwick (Eds.), *Medical enhancement and posthumanity* (pp. 107–136). Springer.

Braidotti, R. (2013). *The posthuman.* Polity Press.

Cassirer, E. (1948). *The renaissance philosophy of man.* Chicago University Press.

Comte, A. (1996). *Philosophie des sciences.* Gallimard.

de Condorcet, M. (1955). *Sketch for a historical picture of the progress of the human mind.* Weidenfeld & Nicolson.

Davies, T. (1997). *Humanism.* Routledge.

Deitch J. 1992. *Post human: Exhibit catalog essay 1992.* FAE Musée d'Art Contemporain Pully.

Dupuy, J.-P. (2011). Cybernetics is antihumanism: Advanced technologies and the rebellion against the human condition. In G. R. Hansell & W. J. Grassie (Eds.), *H+ transhumanism and its critics* (pp. 101–110). Metanexus Institute.

Haraway, D. J. (1991). A cyborg manifesto: Science, technology, and socialist feminism in the late twentieth century. In *Simians, cyborgs and women: The reinvention of human nature.* Routledge.

Hassan, I. (1977). Prometheus as performer: Toward a posthumanist culture? In M. Benamou & C. Caramello (Eds.), *Performance in postmodern culture* (pp. 201–217). Coda Press.

Hauskeller, M. (2013). *Better humans: Understanding the enhancement project.* Acumen.

Hauskeller, M. (2016). *Mythologies of transhumanism.* Palgrave Macmillan.

Hayles, N. K. (1999). *How we became posthuman: Virtual bodies in cybernetics, literature, and informatics.* University of Chicago Press.

Kant, I. (2006). *Anthropology from a pragmatic point of view.* Cambridge University Press.

Mehlman, M. J. (2000). The law of above averages: Leveling the new genetic enhancement playing field. *Iowa Law Review, 85*, 517–593.

More, M. (2013a). The philosophy of transhumanism. In M. More & N. Vita-More (Eds.), *The transhumanist reader* (pp. 3–17). Wiley-Blackwell.

Nayar, P. K. (2013). *Posthumanism.* Polity Press.

Kurzweil, R. (2005). *The singularity is near*. Penguin.
Ranisch, R., & Sorgner, S. L. (2014). *Post- and transhumanism: An introduction.* Peter Lang.
Soper, K. (1986). *Humanism and anti-humanism.* Hutchinson.
Spinoza, B. (1985). *Ethics*. In *The collected writings of Spinoza* (2 vols., Vol. 1). Princeton University Press.
Tirosh-Samuelson, H. (2014). Religion. In R. Ranisch & S. L. Sorgner (Eds.), *Post- and transhumanism: An introduction* (pp. 49–71). Peter Lang.
Vita-More, N. (2019a). History of transhumanism. In N. Lee (Ed.), *The transhumanism handbook* (pp. 49–61). Springer.
Wolfe, C. (2010). *What is post-humanism?* University of Minnesota Press.

4

Theory and Practice of Transhumanism

4.1 Enhancement

The central theme of transhumanism and the concept around which the entire project revolves is enhancement or, more precisely, human enhancement. To sum up what has been said so far, transhumanism aims to make use of any morally legitimate and practically feasible means, above all nanotechnology, biotechnology, information technology and cognitive science (NBIC) applications, capable of upgrading or enhancing the human being. However, if transhumanism were merely this, why should it be the object of so much criticism, doubt, perplexity and dissent? Why criticise or feel concerned about the purpose of practically all human actions? After all, has humanity not always and only tried to improve its own condition and situation since it first appeared on the face of the Earth?

These are challenging questions to which there are no clear and definitive answers but which require complex and thorough investigation. Provisionally, it can certainly be accepted that humanity has always sought to improve its existence, to eat better, to protect itself better from danger and from the climate, to suffer less and to live longer,

F. P. Adorno, *The Transhumanist Movement*,
https://doi.org/10.1007/978-3-030-82423-5_4

which is what transhumanism proposes. And yet, despite an apparent continuity between these actions/aims and the enhancement proposed by transhumanism, even a superficial glance immediately identifies such specific means and procedures for transhumanist enhancement that raise the suspicion—whether or not it is legitimate is what needs to be ascertained—that there is an unyielding discontinuity between the one and the other. Humanity may well have always striven to improve itself, but we have reached a point where enhancement—improvement, optimisation, upgrading, perfection—has become the watchword, the motto of the contemporary world, if not its imperative. The not entirely innocuous consequence of this is that prescriptions are being modelled on a description: since humanity has always improved itself, there is nothing wrong in its continuing to do so. In view of this, it seems essential to understand what these enhancement practices constituting the stock-in-trade, so to speak, of transhumanism actually are. Specifically, is the human enhancement proposed by transhumanists analogous, in terms of its definition, means, practices and aims, to the enhancement or improvement that the human being has always tried to achieve?

Perhaps an appropriate starting point would be to define enhancement itself or, rather, the sense in which this term has so far been used. The first significant appearance of a reflection on enhancement dates back to a famous lecture given by the bioethicist LeRoy Walters at the Hastings Center in 1993, which subsequently led to the publication of a book reporting the collected results of a project carried out by the same institute to investigate the theoretical and practical aspects of human enhancement practices. At this conference, Walters examined the moral legitimacy of four possible and foreseeable interventions of genetic enhancement. He claimed that gene therapy could be used to improve the immune system, decrease the need to sleep, increase memory and reduce human aggression while enhancing altruistic feelings (Parens, 1998a). Irrespective of the identification of the functions selected for enhancement, Erik Parens—the editor of the book published following Walters's controversial intervention—pointed out that the term (and therefore the practices it aimed to define) was both vague and clear at the same time, and for this reason suggested abandoning its use (Parens,

1998b, 2), even though he immediately added that its replacement with another term would not solve the problem.

This seems to fully legitimise the position of authors like John Harris who refuse to give it a specific meaning. According to Harris there is no need for such a definition because any modification that produces a change for the better is clearly an enhancement (Harris, 2007, 36). However, aware of the vagueness of this definition, he promptly points out that we cannot actually know if a modification is an enhancement because we cannot know in advance the situations in which the modified individuals will find themselves and hence we cannot know whether or not such changes will benefit them (Harris, 2007, 36). Savulescu and Bostrom express the same opinion, although for different reasons, in their introduction to the collective volume *Human Enhancement*. They suggest it would be counter-productive to give a precise biomedical definition of enhancement and it would be much better to proceed on a case-by-case basis without bothering with universal and general definitions that would then have to be adapted to specific cases or might pose greater theoretical obstacles (Bostrom & Savulescu, 2009, 3–4). Leaving aside what is to be enhanced and how this is to be achieved, Allen Buchanan suggests a definition that is absolutely generic and somehow also implicit in Walters's reflections: "a biomedical enhancement is a deliberate intervention, applying biomedical science, which aims to improve an existing capacity that most or all normal human beings typically have, or to create a new capacity, by acting directly on the body or brain" (Buchanan, 2011b, 23). Although Buchanan's definition makes explicit reference to a specific type of enhancement (biomedical enhancement), it is so generic as to be of practically no use, and it encounters the same difficulties as those of Harris, Bostrom and Savulescu. On the one hand, it is not possible to know whether a change is always an enhancement regardless of the context of the person subjected to the intervention, and on the other hand, the introduction of the term "biomedical enhancement" probably makes the issue more complex. It is true that extending the range of sounds that a human being can pick up is certainly an augmentation, but can it be considered an enhancement, a change for the better and, as such, desirable? Would it not be necessary to take into account the context in which this new faculty would be used so as to determine whether it is a good idea to augment it and by how much? By following Buchanan, it becomes impossible to distinguish between interventions that are not beneficial to the individual and those that enhance faculties by producing beneficial results.

This definition of enhancement cannot in the least explain the presumed specificity of transhumanism and, therefore, cannot facilitate a more precise evaluation. It might simply lead to the conclusion that, as humanity became aware of its imperfections early on, it is in the order of things that it has been constantly trying to improve itself since then. Transhumanism should not incite disbelief because it only proposes what humanity has always attempted to do: all technological inventions have this purpose and the very function of technology is to enhance life. Indeed, one of the most ardent defenders of enhancement, Allen Buchanan, argues that a careful analysis of the development of the human species would show that enhancement in and of itself is absolutely nothing new and, in fact, that the two are so intertwined as to be inseparable (Buchanan, 2011b, 24). But if this is so, why question the very notion of enhancement which, according to Buchanan, coincides with the history of humanity? Is such criticism, and ultimately its very use, not redundant?

The fact is that human enhancement can refer to several things, and first of all to practices and objects that are quite commonplace: glasses, bicycles and computers are just some examples of means that help to improve the quality of everyday life. On the other hand, there are more "exotic" forms of enhancement, such as becoming cyborgs, having nanobots circulating in the body and uploading our minds, which border on science fiction and are described in minute detail in the works of transhumanists (Coeckelbergh, 2013a, 15). This difference in the practices and aims of enhancement is well summarised by Hauskeller, who defines four general stages: "(1) illusionism, (2) fortification, (3) replacement, and (4) displacement". The first two correspond to widely accepted and traditional practices: surgery or prosthetics aimed at enhancing physical appearance or physical enhancement in the broad sense of the term. The third and fourth are enhancement practices that profoundly alter the human body: what Agar calls radical enhancement. The third achieves enhancement by replacing biological organs with stronger and more efficient artificial organs, the fourth goes as far as to replace the body with another medium capable of performing the same functions. The last two enhancement practices are now proposed by transhumanism, but they had never previously been taken into consideration due to the obvious technical difficulties they entailed (Hauskeller, 2016, 65).

Laura Cabrera is of the same opinion and defines three types of enhancement. "Biomedical enhancement" is "any intervention aimed primarily at the improvement of one or more core capacities of an individual beyond what is therapeutically necessary to sustain or restore health as conceived within biomedicine" (Cabrera, 2015, 38). "Transhumanist enhancement" identifies "any intervention, not necessarily medical, aimed primarily at the improvement of one or more core capacities of an individual beyond *species-typical limits* with the aim of *overcoming* human biological limitations" (Cabrera, 2015, 39). Finally, "social enhancement" is "any intervention that augments or improves an individual's *capabilities set* with the aim to *enable* and *empower* them as *active members of society*, without directly changing the biological reality of individuals". The latter form of enhancement does not directly target individuals or their capabilities, but the environment in which they live. Cabrera claims that people are not isolated atoms but evolve in a context that determines their possibilities and activities; hence, any disabilities they may have are more the result of interaction with their environment than of individual deficiencies (Cabrera, 2015, 93). Although undeniably interesting, Cabrera's analyses seem to provide no definitive indications. The differences between biomedical and transhumanist enhancement are so minute as to infer either that all biomedical enhancements are also compatible with transhumanism or that transhumanist enhancements have the same purpose as biomedical enhancements, that is to produce a situation that is "better than well". The substantive difference may actually be between biomedical and transhumanist enhancement on the one hand and social enhancement on the other: while the former intervenes in the human body, the latter modifies the individual's environment. This distinction will be discussed more fully below.

For the time being, the only result emerging from these definitions is that all human actions are enhancements, thus rendering meaningless any reflection on enhancement and on the transhumanist movement itself. Despite the difficulty of distinguishing between different types of enhancement and, above all, of finding a definition that captures the essence of transhumanist enhancement, the unease that such interventions engender does not seem to be decreasing. Quite the reverse appears to be the case, as Cabrera observes, with the morality of transhumanist enhancement being questioned while biomedical therapy is accepted, even though they cannot be clearly separated from each other (Cabrera, 2015, 64–65).

The reason for this unease might be explained by the confusion surrounding enhancement which, as already stated, may include somewhat inhomogeneous activities that cannot be clearly differentiated. According to Baertschi, *enhancement* entails two inhomogeneous practices: one that aims to transcend normative limits—a characteristic that is generally overlooked—and thus generates nightmares and fears that are not entirely unmerited, and another that makes improvement an essential characteristic of the individual's moral development. While the first type of enhancement (transcendence) is based on the desire to become something better than what one actually is (better performing, more adequate and, in any case, different, and so on), the second type (optimisation) is the will to become the person one wants to be, the accomplishment of one's ideal of life, using all available capabilities (Baertschi, 2009, 82). Transcendence can be said to correspond to the quantitative notion of progress, roughly dating back to Bacon, while optimisation corresponds to a qualitative and much older conception of progress. In short, while optimisation is seen as a centuries-old practice that is somehow coextensive with human nature and therefore acceptable, transcendence is regarded as a highly problematic product of modernity.

Unfortunately, interesting though it is, Baertschi's approach encounters the same difficulties as those seen so far. The first problem is evidently that an individual may desire to fully become the person they are and hence consider it necessary to develop and enhance skills and performance levels: how should the desired enhancement be connoted in this case? As transcendence and hubris, or as optimisation and self-fulfilment? Although great care is required in distinguishing these two aims of enhancement, as Baertschi acknowledges (Baertschi, 2009, 83), this does not seem to be sufficient on its own. Choosing how to define the desired type of enhancement entails a prior determination of what human nature actually is: such a definition will make it possible at a later stage to apply the difference between transcendence and optimisation. While Baertschi clearly indicates that it is entirely justified and fitting to strive towards self-fulfilment and optimisation but inappropriate to seek to go beyond one's nature through transcendence, this criterion remains vague until a valid definition of human nature has been established. It is equally difficult to distinguish between human abilities that are worthy of being enhanced—and how to achieve this—and those that are not. Baertschi retorts with the observation that this difference can be determined on the basis of the adopted concept of "a good life" (Baertschi, 2009, 86). However, Baertschi does not seem to take into account the fact that contemporary society and its values are polytheistic, with the result that all conceptions of a good life are equivalent (cf. Engelhardt, 1991, 61) and therefore everyone has the right to choose the means that seem most appropriate to them in order to achieve their own life plan. In other words, the concept of a good life cannot be adopted without using a metaphysical or ontological criterion that is currently unavailable.

Nevertheless, Baertschi proposes three criteria for choosing an appropriate conception of a good life. First, a good life is embedded in ideals of altruism and not selfishness; second, the enhancement (or optimisation) of a faculty must be the result of effort and therefore be deserved; finally, the result produced by this effort must contribute to the development of the individual's personality, as self-creation is one of the things people hold most dear (Baertschi, 2009, 93). There is an obvious objection to the first criterion: if the means to ontologically establish a morality system are not available, a good life centred on altruism could be countered with a concept of a good life characterised by autonomy, freedom, caring or even selfishness. In other words, it is essential to know the value of value; and an ontological or metaphysical framework establishing first absolute values is needed to avoid falling into an infinite regress. The second criterion also meets with decisive objections. How can remedying innate deficits or enhancing faculties that prevent a fitting self-fulfilment be the result of hubris? Does the goal, that is the attempt to lead a good life, not overshadow the means by which it is to be achieved?

The third criterion, which is effectively a clarification of the second and is accordingly subjected to criticism as decisive as that directed at the first two, contains a different element, namely the valorisation of autonomy. By pointing out that merit is acquired through personal contribution (i.e. an individual is the author of any enhancement and is deserving of it because of how they have used their natural abilities, regardless of any underlying natural talents), Baertschi highlights how this is linked to the concept of a good life as the creation of self. It is now clear that this criterion cannot sidestep the difficulties present in differentiating between different forms of enhancement: individuals remain the authors of their life whatever the means chosen to achieve it. What is the difference between a person who follows a path that leads to their seeking pharmacological support in order to eliminate or enhance performance in the social sphere (for instance shyness, which may have pathological manifestations such as to make everyday life particularly problematic) or in sport, and another person who uses yoga to improve suppleness and, ultimately, physical ability? The obvious and immediate answer is that there is no difference because both are trying to give a certain form to their lives by the most appropriate means. But it seems that the type of means used is a cause for concern, which will be discussed in the continuation of this work. However, unlike other authors, Baertschi can at least be accorded the merit of having drawn attention to the question of autonomy and the need to valorise it in every way possible.

4.2 The Aims of Enhancement

What is the purpose of enhancement? What do those who support and justify it hope to achieve? What moral values does it seek to promote?

Julian Savulescu, Anders Sandberg and Guy Kahane have defined enhancement in a welfarist perspective as "any change in the biology or psychology of a person which increases the chances of leading a good life in the relevant set of circumstances" (Savulescu et al., 2011, 9).

These authors claim that this definition identifies enhancement as an act that raises the value of a life, since changing a person increases the chances of them leading a good life. Each of these practices aims to make people better, that is to increase their wellbeing and to improve what each individual believes is necessary to achieve it. What makes enhancement practices attractive is that they are tools which can or should improve a person's quality of life. When asked to characterise this good and better life and its differences compared to lives in which enhancement is not envisaged, Bostrom merely indicates that it is a totally subjective ideal on which, he believes, transhumanism does not have much to say. Everyone has their own conception of what their wellbeing could or should be, and, whether this conception is right or wrong, transhumanism simply provides the means to achieve it. Bostrom cannot resist the temptation to mention the view that the wellbeing envisaged and promoted by transhumanists is proportionate to a quantitative increase in their abilities, much like Mill's classic thesis that a good, happy life is a life that contains a certain amount of higher pleasures (Bostrom, 2009a, 111). A good life from the transhumanist perspective would therefore be a life in which cognitively complex and/or pleasure-giving activities are appreciated, understood and pursued.

For their part, Savulescu, Sandberg and Kahane set out a similar argument. Although everyone has the right to achieve their own conception of a good life and to seek this individual fulfilment autonomously and independently, there are certain factors that must be present and in a sufficient quantity (Savulescu et al., 2011). On the basis of common sense, the authors conclude that a high level of intelligence (the only topic they tackle) is an essential factor for human wellbeing. This rather trivial observation is corroborated by the fact that everyone tries to enhance their cognitive abilities in every possible way and with every means at their disposal, while simultaneously staving off any intellectual deficits for themselves and for those close to them.

The key feature of these considerations is the direct and causal link between happiness/wellbeing and control: people will be happier because the enhancements advocated by transhumanism will make them freer by increasing the control they have over themselves, as Hughes, Bostrom, Sandberg and others claim. Transhumanism is seen as a movement of progressive liberation. Sandberg is extremely clear on this point: not only does every individual have the fundamental right to live and to be happy, but also the right to be free, which is consequential on the first two. Freedom is, in turn, manifested in the full possession of one's own body, which makes any desired change justifiable in the search for and attainment of happiness (Sandberg, 2003, 57).

Under the assumption that freedom is a fundamental right, Sandberg argues that it also entails the right for anyone to do what they want with their body. It would be paradoxical to claim that one is free while having restrictions imposed on the use one can make of one's own body, which would appear to constitute the central and fundamental feature of freedom. Furthermore, the rights one has over one's body cannot but include the right to its form, which can be extended to the right to transform it or to oppose any unwanted or imposed transformation. In this regard, the theory embraced by transhumanism does not appear particularly revolutionary; on the contrary, the freedom it advocates is completely coherent and compatible with Western societies' propensity for self-fulfilment, that is the possibility to choose what one is and the freedom to fully express one's subjectivity, starting from the shape of one's body.

This freedom from interference in the private sphere immediately translates into increasing control over one's own subjectivity, even at a biological level. If everyone is free to choose what they want to be, then they must have full control over their own biological processes, which could otherwise hinder the achievement of their life plan, whatever that may be. Moreover, as morphological freedom is indeed a process of self-acquisition, it is also a manifestation of that freedom, that is the expression of one's own subjectivity freed from the limitations of uncontrollable obstacles. Once an individual is fully autonomous and independent of the constraints that every living person suffers and that hinder full control over oneself, one can manifest one's own self and one's own authenticity in every conceivable and desirable way. In other words, full morphological freedom assures the possibility for the individual to construct and manifest their true or authentic personality (cf. Elliott, 2004; Pessina, 2016; Rothman & Rothman, 2003).

In summary, transhumanists defend both positive and negative concepts of freedom (Berlin, 2002): both must be protected and enhanced. Transhumanism aims to make humanity freer, in the sense of negative liberty, because it will remove all biological obstacles, that is the constraints inherent in the living constitution of the human body that prevent its full deployment. On the other hand, it also aims to set humanity free in the positive sense of the term, since the elimination of the obstacles constituted by biological weaknesses and imperfections will enable it to live a truly good life, on which it puts forward clear and precise indications. Some transhumanists, however, neither explain nor take into account this difference between positive and negative liberty and fall into rather contorted forms of paternalism (Harris, 2007) and may even display a certain propensity to moral and political totalitarianism, which Berlin himself warned against in making the distinction between positive and negative liberty.

It could be objected that this unlimited freedom (ultimately a form of discretion with no specific purpose other than to indefinitely change one's form), which Hegel identified as a bad infinity, is a senseless and thus alienating movement. This view is reinforced by the role that technology plays in this expressive research conducted at one's own risk on one's own body. According to Sandberg, there is a constant reciprocal pursuit of technology and morphological freedom, in that technology enables new forms of creation and self-expression which, in turn, look to technology in the expectation that further creative spaces will be opened up (Sandberg, 2003, 58). Thus, it is no longer clear whether technology is the driving force behind morphological freedom (hence diminishing its spontaneity) or whether morphological freedom is excessively focused on the potential of technology, considering this alone as an expressive tool of individual subjectivity.

Clearly, transhumanism sees itself as a movement of liberation and emancipation from external power and control over people's lives. Whether it is a question of gaining freedom from the yoke and sovereignty of biology over individual lives rather than political dominion exercised by a tyrant or moral control exercised by society seems to make very little difference: according to transhumanists, human beings are subject to their biology and can only be liberated by breaking the biological chains that keep them enslaved. This is what Bostrom asserts in the previously cited passage (Bostrom, 2005c, 87): if transhumanism proposes "a greater degree of control over our own lives", then we do not control ourselves, we are controlled by external forces and we are not autonomous.

In light of the above, might it be reasonable to regard transhumanists as liberals? Despite a superficial analogy, the moral and/or personal autonomy claimed by Kant and Mill is very different from the biological autonomy pursued and sought by transhumanists. For Kant, freedom consists of the ability to be one's own legislator, for Mill it lies in the ability to eliminate external influences, while for Bostrom it entails the ability to enhance and increase control over one's own biological substance. Moral and biological self-determination have goals that are not identical. If Kant and Mill see moral self-determination as the fundamental value, it is because the individual who is his or her own legislator is not subject to a law imposed from outside. Moral self-determination therefore has a purpose which is certainly an interminable work in progress, but which remains a regulating principle of the path to full moral autonomy. Biological self-determination appears to be different. The elimination or reduction of the negative effects produced by individuals' biological weaknesses is geared towards a quantitative and qualitative performance enhancement, albeit in a specific or purely aesthetic context. For instance, greater control over one's own body, that is the achievement of morphological freedom, may make it possible to enhance performance in certain physical and/or cognitive areas without a precise, overarching reason for this autonomy being specified; whereas changing one's physical appearance through aesthetic surgery does not require any justification or legitimation, but merely the desire to change. While moral freedom engages the subject in the world and is exercised in relation to others, morphological freedom accentuates the subject's withdrawal within themselves and shrouds them in a narcissistic and indefinite concern for their own identity.

Secondly, while moral control over oneself can be pursued in total autonomy and solitude through an effort of will and reason, control over one's own biological substance, as manifested in cognitive or moral performance enhancement, must entail direct intervention on the individual body. There is thus a third party, comprised of biotechnologies and their operators, whose impact on the ability to control oneself and to be in possession of oneself cannot be ignored. One might even wonder whether the constant presence of a third party—the technologies required to match this increase in morphological freedom—does not dramatically reduce the ability to assure the increased control over oneself sought by Bostrom.

Let us consider the case of two people who display aggressive and socially condemned behaviour. The first commits to psychoanalysis or simply strives to exercise as much control as possible over their violent impulses, and then tries to improve or control themselves in traditional ways, or in any case not using biotechnological or pharmacological means because, among other things, they do not perceive this problem as a pathological disorder. The second person, who perceives their disorder as pathological, relies on a lithium-based therapy which, although not particularly innovative, shares its pharmacological nature with transhumanist enhancement. The first person gains autonomy because they do not have to seek outside help in order to optimise their behaviour, thus achieving two results: they improve by eliminating an unwelcome behavioural trait and learn how to deal with similar situations without resorting to outside help. The second person, on the other hand, passes from one addiction to another, locked in a vicious cycle of debilitating aggressiveness and the use of substances over which they have no control. Clearly, there is no real upside in this scenario, only displacement: the subject does not become more autonomous, but merely shifts the source of their dependence, passing from a lack of autonomy caused by internal factors to a dependence on external factors. It goes without saying that suspending lithium would simply take them back to the starting point. This trivial example confirms what C. S. Lewis argued in the mid-twentieth century in his analysis of the idea that humanity has conquered nature: in fact, it is nature that has definitively conquered humanity. The use of technology is not liberating but, on the contrary, adds another dimension to the servitude that already oppresses people. First, new technologies do not augment people's power but diminish it while increasing their dependency. This is because they use a medium that is always on loan (or whose use is always subject to external factors, such as its production, the user's lack of ownership, and so on). This means that just as technology enables people to do more, its disappearance might take their control below the threshold that was in place before technology raised it. Second, if nature is all that humanity can control, then all that can be controlled is nature; if humanity can increase control over itself by discovering and manipulating its biology to an ever higher degree, then it follows that it does not abstract itself from its biological conditions, though the latter become pervasive (Hauskeller, 2016, 89; Lewis, 1943, 72–73).

Specifically, it follows from the above that transhumanism places the accent first of all on the *materiality* of the human being, on the imperfection of his biological form. A second aspect is the importance attributed to *direct intervention techniques* to modify the characteristic features of the human body. Thirdly, these techniques have *a common and precise purpose*: enhancement, an increase in the performance of the human body. Fourthly, the aim is *to achieve greater control over the biological processes of individuals*, which is considered as a bringer of greater freedom. Transhumanism is therefore a theory that maintains that *the innate imperfection of the human body justifies and legitimises direct intervention in function of its "enhancement" which has the moral purpose of improving its wellbeing through an increase in individual freedom.* In other words, as is evident from the definitions provided here, transhumanism is a philosophy of life which theorises the enhancement of certain biological features of humanity through the use of biotechnology, in the belief that these changes will make it happier because they will increase its freedom. This rather vague definition, which will be examined in greater detail below, corresponds quite precisely to what transhumanists have proposed and which still poses important theoretical problems that will be analysed hereafter.

4.3 What Needs to Be Enhanced?

Bostrom states that there are at least three fields in which enhancement would be particularly important: *healthspan*—the capacity to remain fully healthy, active and productive, both mentally and physically; *cognition*—general intellectual capacities, such as memory, deductive and analogical reasoning, and attention, as well as special faculties such as the capacity to understand and appreciate music, humour, eroticism, narration, spirituality, mathematics, etc.; and *emotion*—the capacity to enjoy life and to respond with appropriate affect to life situations and other people (Bostrom, 2009a, 108). The enhancement of at least one of these basic faculties beyond the "maximum attainable by any current human being without recourse to new technological means" (Bostrom, 2009a, 108) defines the individual as *posthuman.* Bostrom acknowledges that other human faculties and characteristics can also be improved and enhanced, and indeed a more precise and detailed list taking into account the literature on human enhancement would also include enhanced physical performance and the modification and modulation of moral evaluations. In short, enhancements can encompass physical performance (body enhancement), cognitive abilities (cognitive enhancement), mood, emotions and feelings (mood enhancement), morality (moral enhancement) and the extension of individual life resulting from a general improvement in health.

These five capabilities do not all have the same weight and importance: the extension of human life is clearly the primary objective of enhancement techniques. To avoid becoming melded into the rhetoric used in the specific literature, however, it must be made clear that it is not simply an increase in the average life expectancy of individuals, or an extension of human life expectancy that is being discussed, but much more directly and frankly the achievement of immortality considered as "the Holy Grail of enhancement" (Harris, 2007, 59) since, as is claimed in the context of transhumanism, vanquishing death is the supreme goal (Tirosh-Samuelson, 2011b, 16). The fundamental purpose of transhumanism is to make human beings immortal, that is to bring about the "death of death", the pure and simple elimination of what is considered the absolute evil (Alexandre, 2011; cf. Hauskeller, 2013, 118; More, 1996).

From this it follows, first of all, that the length of life is good in itself: for transhumanists, as long a life as possible spent in good health and to the fullest extent of one's abilities is something good that should be pursued and preserved. In actual fact, the pursued aim should be to arrest human life in a sort of biological immaturity rather than to extend it. Biological immaturity is the condition in which all potential is on the verge of coming to fruition and the individual is still indeterminate: everything is still possible, and no alternative has yet been discarded through the fulfilment of some potential. However, stabilisation in this extremely dynamic state is not contradicted by actualisation, which is always provisional and liable to be overturned: transhumanism asserts that all potential can and must be fulfilled. In the infinite time provided by immortality, there would be no reason why this should not happen.

Each of these enhancements presents certain critical elements intrinsic to the notion of enhancement itself, in addition to some peculiar problems linked to their specificity. For example, as far as moral enhancement is concerned, it is difficult to identify a human organ tasked with the formulation of moral judgements, while it is obviously straightforward in the case of enhanced physical performance. It is equally clear, however, that the enhancement of any objectified organic functions cannot be defined except in relation to a specific purpose and according to a means–end scheme identical for all its forms.

4.4 Body Enhancement

Body enhancement may serve two purposes: *functional alteration* and *formal modification*, which very often (though not always) are a consequence one of the other (Devereaux, 2008; Gordijn, 2006; Jones & Whitaker, 2012). Apart from the differences in the aims that body enhancement may pursue, both are supported and justified by the idea that all individuals enjoy full morphological freedom, which is thus the essential condition for the achievement of these two objectives, one aesthetic and the other functional. Let us begin with the first, that is the will to modify one's body for aesthetic and artistic purposes. In this regard, a transhumanist aesthetic follows (or may more or less explicitly follow) two different paths.

The first path seeks to make the human body conform to a socially accepted ideal of beauty primarily, but not exclusively, through cosmetic surgery. Modification of the body is thus an enhancement to the extent that it allows an individual to feel comfortable by conforming to predominant values. However, despite the extensive literature on this issue, transhumanism does not appear to place social conformism achieved through physical enhancement among its main values. On the contrary, it seems that Sandberg's precious morphological freedom goes far beyond the aesthetic conformism that leads so many people to undergo cosmetic surgery (Elliott, 2004; Rothman & Rothman, 2003), since it merely indicates that everyone has the freedom to do what they want with their body, using all and any means that seem to make their goals achievable (Miah, 2011).

The second path sees the body as matter to be modified, essentially for artistic purposes. In this context, the transhumanist aesthetic does not seem to be particularly innovative: the human body has always been, if not the preferred object, certainly one of the most frequently represented themes in art. However, the practices of hybridisation, of body art from tattooing to scarification, also confirm the importance of the body as the privileged medium of artistic practices aimed at modifying it or, at least, at utilising it as the medium for its representation. As Natasha Vita-More argues, transhumanism reflects nothing but a non-fixist conception of the human being, which has long been at the centre of artistic and aesthetic reflection (Vita-More, 2013, 24). Transhumanism proposes nothing but an indefinite extension of humanity's aesthetic capacities and artistic sensitivity through the application of biotechnologies to his body and his immersion in a constantly shifting space and time with no further specification given to the meaning of this artistic operation. In an intense passage (quoted in full below) Natasha Vita-More explains what transhumanist art and aesthetics should consist of:

The modification of biological life systems, from single cells to organisms, increases the transdisciplinarity of the arts and sciences. As noted, some practices have reached far into the uncomfortable zone of bioengineering and genetics, where science and medicine reside, in aptly creating bio-experiments and offering opinions on the meaning of life. On another side of the creative spectrum, exploratory creations with nanotechnological robots have become a molecular vehicle for establishing artistic practice and theory. The transhumanist art arises when we combine biodesign and nanodesign, along with information technology and cognitive/neuro science for life expansion, converging nanotechnology, biotechnology, information technology, and artificial general intelligence to provide the transdisciplinary media for investigating the continuation of life by enhancing, extending, and regenerating life in biological, synthetic, and cybernetic forms (Vita-More, 2013, 23).

In short, it is a question of experimenting with creativity in order to imagine and produce unknown forms of the human being, to create biosculptures of which each individual is both the producer and the material. Transhumanist aesthetics is imprinted with vitalism as an end in itself, whose aim is to produce a more or less artificial form of biological life (Delfin, 2019; cf. Critical Art Ensemble, 1998; Herath et al., 2016).

Alongside the manifestation of its own morphological freedom with purely aesthetic and artistic goals, transhumanism also supports the use of biotechnology aimed at improving the quality of life through the enhancement of physical performance. If the physiological modification of one's own body through an increase in physical performance can be justified by the fragility and incapacity of the individual, by the positioning of the ends sought beyond one's capabilities or by the existence of social competition to achieve them, then in reality the ultimate foundation always lies in the freedom to do what one wants with what one is or what one possesses or, more precisely, with one's own body. Here too, transhumanism does not seem to raise unheard of or particularly new moral issues.

The body can be enhanced to improve, or enable when lacking, those functions that are considered statistically specific to the individual. Broadly speaking, body enhancement aims to raise certain functions and performance levels above an individual's threshold level: it can therefore allow those who possess abilities below the statistically observed threshold of normality for that function to reach or exceed this threshold. As normality is established statistically, it simply serves as a reference point for measuring improvements, but not as a benchmark for enhancement practices. This does not detract from the fact that in specific contexts, enhancement produces benefits that are considered illegitimate, particularly in professional sport. Indeed, a large part of the transhumanist literature focuses quite repetitively on doping practices that encompass all the problems posed by body enhancement for functional purposes. The improvement of physical performance is an innate tendency of the human being which now, thanks to the change in the theoretical-conceptual structures of modernity and the unprecedented progress of technology and science over the last century, appears potentially unlimited. It would be trite to say that humanity strives to enhance its intellectual and physical performance with every means at its disposal: an obvious example of this is sport, the aim of which has always been to do more and better, ever since the Olympic motto "*Citius, Altius, Fortius*" was coined; the use of enhancing products and strategies to achieve this end has been a constant. The characteristic features of enhancement techniques are very similar: they always entail manipulations aimed at improving the specific functionality of one or more organs. From the moral point of view these practices raise the same objections and present the same criticalities when they are extended to society as a whole. Although there are different forms of doping in sport, it is generally Western society that uses techniques and products aimed at improving human performance in a purely economic vision of existence based on the notion of profit. Implanting a prosthesis to remotely control household appliances would save time and ensure greater speed of movement. The paradox is that the time saved in this way is never actually spent but is reinvested so as to save even more time in an indefinite dynamic that not even death can interrupt. Overall, transhumanism is committed to defending morphological freedom as such, regardless of the use that is made of it, albeit in a utilitarian perspective, since it represents the essential means to increase individual wellbeing, however individuals define it.

4.5 Cognitive Enhancement

In the transhumanist sphere, cognition has been defined as a series of processes used by an individual to organise information. Basically these processes include "acquiring information (perception), selecting (attention), representing (understanding) and retaining (memory) information, and using it to guide behaviour (reasoning and coordination of motor outputs)" (Sandberg, 2011, 71; cf. Bostrom, 2009a, 117; Hildt & Franke, 2013; Housden et al., 2011, 113; Niyadurupola, 2008; Turner & Sahakian, 2008). Cognitive enhancement may entail any of these processes through the use of completely new NBIC means in various forms of prosthetics, magnetic stimulation of specific areas of the brain and direct intervention on the individual's genetic heritage.

Prosthetic grafting envisages two types of intervention: pharmacological and physical. The former is actually quite traditional: caffeine and nicotine are known to increase concentration and stimulate attention by reducing intellectual fatigue, and it cannot be said, at first sight at least, that drugs like Ritalin or Modafinil have qualitatively different effects: they just seem more effective. So why is there such vehement opposition to them? The fundamental reason is that Ritalin and Modafinil, and also other pharmaceutical products that enhance human wellbeing in one form or another, are basically drugs, that is substances that differ from caffeine and nicotine because they are not for normalised, daily, social and hedonistic use but have a therapeutic purpose. Caffeine and theine are taken in the form of drinks in order to stay awake and remain focused, but above all because their consumers find coffee and tea pleasant. Modafinil and Ritalin lack the hedonistic aspect typically associated with the consumption of coffee, tea or cigarettes and they change the individual's perspective on the states they aim to eliminate. In other words, attention deficit disorder (ADHD) or a drop in intellectual concentration become pathological states and not normal moments of people's everyday life. As for physical prosthetic grafting, that is hybridisation with computers or other electronic devices, there seem to be no contraindications for their use. Any issues they raise will be discussed in the following sections. At this stage it is only necessary to point out that their contribution to the enhancement of cognition comes from their direct implementation within the biological matter of the individual, which has seemingly far more radical implications than with other forms of physical prosthetics.

The same problem seems to arise with two other means generally used or advocated for cognitive enhancement, transcranial magnetic stimulation (TMS) and genetic manipulation, and it is clearly the invasiveness of these techniques that arouses a certain mistrust. TMS has proved effective in enhancing many aspects of cognition: it has enhanced performance in a procedural learning task, in motor tasks, in visuo-motor coordination, in working memory, in classification and even declarative memory consolidation during sleep, and finally in some linguistic tasks. However, its long-term effects are unknown, its use requires great attention to individual variables and it also poses important health risks, not least the possibility of producing epileptic seizures (Sandberg, 2011, 75).

As for genetic manipulation, tests have been conducted to assess its efficacy in enhancing memory capacity. However, any resulting progress has so far been fairly insignificant and Sandberg himself acknowledges that the benefits of genetic changes for the enhancement of individual intelligence are almost negligible, while the risks related above all to their intensive use and to their long-term effects appear at first sight to be far more serious.

The question that now needs to be answered is whether cognitive enhancement is truly beneficial and for whom. The first point concerns the improvement in individual wellbeing that can be expected from the use of "cognitive enhancers". Cognitive enhancement would predictably translate at once into a more effective functioning of instrumental rationality as well as a major positive impact on the choice of one's own life plan. In other words, it would be possible not only to offer a fuller specification of the means needed to achieve a goal, but also to understand whether or not the intended goals are worth pursuing. But would this enhancement be sufficient to make people happier? Apparently not, Bostrom argues, because genetic heritage plays such a predominant role that, no matter how much effort is devoted, there would be no significant effect on wellbeing (Bostrom, 2005c). Even if Bostrom's hypothesis should turn out to be wrong, there would still be some doubt surrounding a directly proportional relationship between cognitive enhancement and an increase in individual wellbeing and happiness. Such a relationship might be observed in the correlation between low cognitive abilities and unhappiness rather than in particularly intelligent people (Sandberg & Savulescu, 2011, 96). However, Hauskeller states that if there is a relationship between cognitive ability and happiness, it is inversely proportional: enhanced cognition seems to result in a fall in individual happiness or wellbeing and not in an increase or an improvement. Chesterton claims that poets and artists in general are highly unlikely to go insane, while this is very often the case with chess players or mathematicians: "The madman is not the man who has lost his reason. The madman is the man who has lost everything except his reason" (Hauskeller, 2013, 19–20). Although an increased memory capacity might have an enhancing effect on cognition, it does not seem to contribute to individual wellbeing, unlike the ability to forget. In actual fact, there is no implicit utility in increasing either of these, as it would seem more reasonable to prioritise the relationship between the one and the other, that is it is the balance between the ability to remember and the ability to forget that contributes to individual wellbeing. In this case, however, it is a qualitative relationship—in the sense that there are some events that one prefers to forget and others that one prefers to remember—which is independent of the quantity of events one is capable of remembering or forgetting. The desire to store some memories rather than others does not seem to improve with the use of technologies, which cannot select memories but only regulate their quantity.

It seems not inappropriate to add emotions and affective states generally to this list of functional processes and abilities: as has been observed, mood enhancers, such as Prozac which increase and improve cognition indirectly (Hauskeller, 2013, 14), are not cognitive enhancers in the strict sense of the word but facilitate the use of cognitive faculties.

Bostrom admits that, while it is difficult to characterise physical enhancement and even more difficult to explain how and where to act in order to increase or enhance cognition, it is almost impossible to understand what enhancement of the emotions might entail (Bostrom, 2009a, 111). Presumably, mood enhancement aims to increase control over one's feelings, or more precisely one's emotional fitness, and all affective manifestations, feelings and emotions that make us lose control of our responses to stimuli of any kind should be eliminated. Mood enhancement makes us better insofar as it allows us to consciously manage our affectivity. However, the characteristic feature of emotions, affections and feelings is that they are uncontrollable by definition and, in actual fact, they are what they are precisely because they make us lose control of ourselves by catapulting us out of ourselves and into the unknown. In other words, mood enhancement would result in the elimination of the entire passive aspect of feelings and emotions that characterise them and, ultimately, in their definitive elimination. Mood-altering drugs or therapies promise increased control over oneself, which explains why mood and cognitive enhancers seem to be popular with the public at large (Rose, 2007, 101).

In addition to the individual improvements indicated by Bostrom, emotional enhancement is the result of an entirely singular and individual assessment. It is the construction of an individual character that, as such, is left to people's free choice. The only criterion that can be used to evaluate emotional enhancement is once again the resulting sense of wellbeing. Although mood enhancers may also prove useful in the treatment of certain psychological pathologies, they pose a general problem because it is solely the person and not the context in which they live that is problematic and therefore in need of a remedy or cure (Rose, 2006, 73). The adaptive pressure is placed squarely on the shoulders of the individual, who is asked to adopt behaviours consistent with the environment in which they live: if they are unable to do so, the environmental demands are certainly not called into question but, rather, it is the person who must provide a justification for their alleged pathologies and remedy them.

Leaving aside the technical problems, that is whether or not a drug or technology is actually able to enhance cognition, the efficacy of cognitive enhancers poses a theoretical problem in that it is unclear what knowledge enhancement really means, just as it is unclear what enhancement of the mood and emotions means. According to Bostrom, these moral conundrums stem more from the fact that they are new techniques than from intrinsic problems (Bostrom, 2009a, 119).

There are, however, some specific critical issues concerning the impact they may have on personal identity, authenticity and personality, that is on character (DeGrazia, 2005; Elliott, 1999, 2004). With regard to the risk that enhancement might lead directly to an authentic manifestation of the self and a true expression of an individual's nature, there may be some doubt over whether transhumanism might actually constitute such a danger and also whether it is a real danger. To further blunt the view that transhumanism poses a threat from this standpoint, it is argued that in order to discuss how mood enhancement might endanger personal identity, there should first of all be a clear and unambiguous notion of the criteria for defining personal identity (Glover, 1988). Only then would it be logically possible to attempt to identify the possible risks that the use of these new technologies could pose to personal identity (Berghman & alii, 2011, 158).

However, the debate would be much broader-based and would call into question the very notion of authenticity, which appears to be a resolutely modern ideal. Authenticity means being in contact with oneself, with a deep and singular part of one's personality that constitutes the essence of uniqueness. On the one hand, this means that authenticity and morality are not necessarily coincident, not because authenticity can be both moral and immoral but because the search for individuality takes place outside the scope of morality and cannot be evaluated in moral terms: the originality and uniqueness that characterise authenticity are not moral characteristics of the individual because they cannot be evaluated using such terms as "right" or "good". On the other hand, the means that can be used for this purpose have no impact on the search for authenticity: if the end cannot be morally characterised, then neither can the means.

Finally, it is paradoxical that the use of biotechnology is considered particularly suitable for reaching or achieving individual singularity, that is for recovering or manifesting one's true self (Elliott, 2004; cf. ter Meulen, 2015, 93). From the transhumanist standpoint, not only do NBICs not endanger people's true nature, but rather they help to recover it or to rediscover it by removing all the obstacles that limit its full manifestation.

4.6 Moral Enhancement

The debate on the legitimacy and necessity of cognitive enhancement has seen some recent developments that, while predictable, have nevertheless raised a few eyebrows (cf. Lavazza & Reichlin, 2019). Not content with analysing the role emotions play in cognition, the supporters of enhancement have applied themselves to defining the characters and techniques of "moral bioenhancement" (MBE), which was originally not a concern of transhumanists.

Starting from its technical feasibility, MBE poses a number of different problems from those encountered in questions of physical or cognitive enhancement, while others are analogous due to the alleged need for a quantitative increase in the functions of a certain organ that produces judgements and moral actions according to an organ–function–performance structure. A successful transfer of this combination of elements to morality seems highly improbable, not least because it would be difficult, if not impossible, to identify the moral organ whose functions need to be enhanced, assuming, of course, that such a moral function exists. What is more, it would be extremely difficult to give a univocal definition of MBE's aims (and much harder than for all other forms of enhancement). What does "morally enhanced" actually mean? More altruistic, more just, more reasonable, more virtuous?

In order to clarify the aims of MBE, it is necessary to initially assume a particular moral perspective, the priority of which should be justified. In actual fact, the MBE debate is less about moral enhancement proper than about how to eliminate the negative impact of emotions, feelings and passions on moral judgements. In short, it is a psychological context (psychology applied to moral judgements) rather than a question of moral philosophy proper. This is the conclusion that emerges from Douglas's first definition of MBE (subsequently reformulated slightly), focused entirely on strengthening the reasons for an action: "A person morally enhances herself if she alters herself in a way that may reasonably be expected to result in her having morally better future motives, taken in sum, than she would otherwise have had" (Douglas, 2008, 229; cf. 2013, 161). If MBE is an attempt to have morally better motivations and to understand them as causal mental states of our actions (Douglas,

2008, 229), then it would be a question of considering a direct intervention in the emotions that condition moral behaviour. This complies with a Humean conception of morality, according to which actions are not produced by reasons or by what is believed to be true, but by the emotions and desires that are the direct causes of those actions.

The aim of MBE might thus be to eliminate or lessen emotions and stimuli that somehow prevent morally right behaviour by interfering with the good reasons to behave correctly and promoting a behaviour that, while not immoral, is certainly not completely moral. The weakening of these negative stimuli, among which Douglas explicitly considers those that lead to racist and aggressive behaviour, would have the consequence of restoring full causal efficacy to the "good" reasons (Douglas, 2008, 231). In order to clarify his thinking in this regard, Douglas uses the example of a "biased judge" who, nourished by racial prejudice, does not judge white and African American defendants with the necessary impartiality but reserves harsher sentencing for the latter (Douglas, 2013, 161). Consistent with the definition of MBE, the moral enhancement of such a judge would entail eliminating his racial prejudices, which have a negative impact on his judgments by making them unequal. Moral enhancement, while based on an implicit normative conception, is therefore not a problem of normative ethics, at least not explicitly, but focuses on the specifically meta-ethical issue of the causal force of our moral motivations.

However, it is unclear how stimuli, motivations and emotions of this kind, and only of this kind, can be weakened, especially as they should first and foremost be identified as the cause of behaviour considered to be wrong or, more correctly, immoral or illegitimate. Consequently, there are two aspects that need to be discussed and which are in any case the object of criticism. The first is that even experts in neurology and cognitive science, such as Damasio, have not attributed emotions and feelings with the role of decision-making agents, but simply of decision accelerators operating on the basis of acquired cognitive elements, as in the case of somatic markers (Damasio, 1994).

Second, it should be noted how indebted this eventuality is to a re-emerging behaviourism and to the reflex arc theory that underpins it. It is not possible here to indicate even briefly the consequences of this implicit reference. Suffice it to say that behaviourism and emotional non-cognitivism are two sides of the same coin (Freedman, 1998) which was criticised with some success by Merleau-Ponty who convincingly demonstrated that behaviours are not responses to nervous stimuli, but arise from a specific assignment of meaning to the world. Consequently, the tools that may effectively modify individual behaviour are those capable of restructuring the subject's world and thus of changing its semantics (Merleau-Ponty, 2014), rather than suppressing negative stimuli or reinforcing positive stimuli.

The idea that moral enhancement can be achieved simply by eliminating the impact of allegedly negative emotions on our behaviour, however tempting its simplicity may appear, fails to take into account the fundamental role played by the cognitive elements that accompany and underpin moral actions or judgements. Moreover, it is argued that morality is not a question of cognition or, to be more precise, that it is not a cognitive problem and, even more drastically, that increasing or enhancing the cognitive tools that biologically pertain to our species would have no impact on our behaviour (Rakić, 2014; cf. Persson & Savulescu, 2012a). According to MBE supporters, the greatest impediment to the execution of moral behaviour is not cognitive but conative, since the human being is reluctant to follow what reason or moral intuitions indicate. The centuries-old dilemma of the weakness of the will lies outside the scope of this work, but it is perhaps worth mentioning that if an individual does not always choose good, it is not because their will encounters obstacles that it does not have the strength to overcome but because, in reality, they choose what actually seems better or what they want more. Knowing where good lies and choosing evil is not the result of a weakness of will or an absence of motivation but, much more simply though strangely enough, of the choice of an "immoral" action considered to be better than another that may seem more morally right at first sight (Leibniz, 2005).

It has been argued that a right action (whatever that means) is not a necessary and legitimate consequence of feeling a certain kind of emotion, or alternatively, it is not necessary to have feelings of goodwill or empathy to act justly. In short: "I don't have to be nice to be good" (Harris, 2013, 172). As Harris observed, the traditional methods of moral enhancement (i.e. not involving any biological intervention and hence ME, not MBE), such as education, act fundamentally on cognition, on the ability to distinguish good from evil and to understand that violence is morally wrong and that respect for others is morally right. Feelings can be taught, and it is possible to learn how to act in a just and fair way by striving to understand events, subjecting behaviour to constant scrutiny and improving habits (Harris, 2011).

Douglas responded to these objections by pointing out that some commonly practised ME procedures which eliminate negative stimuli or reinforce positive stimuli mechanically are non-cognitive and do not entail an enhancement of cognition and its tools (Douglas, 2013). Douglas cites the case of a sex-addict "driven" by "unwanted sexual desires" (and one wonders whether his sexual desires are indeed unwanted and what it might mean to have a will so weak as to be manipulated by desire), whose behaviour leads him to be disrespectful of the people who love him and are close to him. According to Douglas, one way he could achieve moral enhancement without resorting to anything cognitive would be to avoid finding himself in situations that put him at the mercy of his uncontrollable sexual desires, or to increase the cost that the search for new partners would impose on his personal life (Douglas, 2013). These are unconvincing examples, claims Harris, who nonetheless discerns cognitive processes therein, such as the ability to link causes and effects and understand what causes pain and suffering to others as well as the intelligence to study situations where there are no temptations or that involve excessive risks for the personal life of the sex addict, which are the very elements on which any traditional education founded on rational work is based (Harris, 2013, 169–170). In short, even when it seems that there is no place for cognition, human actions are guided by and based on cognitive elements.

In their criticism of the separation of morality and cognition, Carter and Gordon raise other interesting objections that partially overlap with what Harris observed. First, it is somewhat contradictory to separate cognitive and moral enhancement: it is unclear how progress can be made in the latter—which is necessary, given the current level of our cognition—without somehow increasing and enhancing the former as well, thereby creating a situation that can be labelled as "damned if you do, damned if you don't". Second, it has been observed that the cognitive dimension is an essential element of moral enhancement because the idea that one can live well without knowing (or without trying to know) what it means to live well is utterly contradictory: practical reason needs some form of support from theoretical reason. Third, considering emotions as having only a conative function is an idea that seems at least doubtful. Emotions are also tools of knowledge: improving their application is to optimise the cognitive aspect (Carter & Gordon, 2015; Nussbaum, 2001).

More recently, Persson and Savulescu stated that MBE should aim to increase or enhance altruism and extend a sense of justice.[1] Drug therapy is not intended to *replace* traditional techniques for the moralisation of humanity or absolutely fundamental political changes, but to *accelerate* their effects. However, the context of this reflection does not seem to be different from Douglas's, as can be deduced from the pharmacological therapies indicated: oxytocin and SSRIs. Oxytocin is a hormone and neurotransmitter produced naturally by the human body that seems to have effects on socialisation. According to Persson and Savulescu, this hormone reinforces care-giving and caring behaviours and increases feelings of empathy and generosity towards others: its "cuddle hormone" nickname seems entirely appropriate. The level of oxytocin can be increased by the combined administration of different products, such as certain types of contraceptives or glucocorticoids (Persson & Savulescu, 2012a, 118), whereas SSRIs are molecules that inhibit the absorption of another neurotransmitter, serotonin, and this increases individuals' willingness to cooperate, although its efficacy is disputed (Churchland, 2012).

Even if they prove to be effective in achieving the expected results, it is not certain that they will actually result in moral enhancement. For

[1] Tom Beauchamp (2015) expressed a strong perplexity on the possibility of giving a correct picture of MBE, which stems from the theoretical weaknesses of Persson and Savulescu's work. Persson and Savulescu clearly state that the central aim of MBE is to enhance altruism and a sense of justice, as Beauchamp observes, although this choice is not justified by any argument, nor is it clearly specified which sense is assigned to empathy and a sense of justice nor, above all, is there an indication of what the indispensable level of these two qualities is in order to regard MBE as having been achieved. The latter observation is decisive as the need for MBE is justified by an alleged supposed deficiency of these "functions".

example, oxytocin's ability to raise a person's level of trust in others is not necessarily a plus. It certainly cannot be said that an individual's increased trust in someone who has already deceived them will be beneficial (Persson & Savulescu, 2012a; cf. Baumgartner et al., 2008). SSRIs tend to increase an individual's willingness to cooperate, but this inclination might well prove harmful as any intervention in the stimuli to act does not take into account the individual's situation. It may well be true that cooperation is generally an aspect of morality, but this inclination becomes positive (or advantageous) and morally just when the context is taken into account. Ultimately, intervention in an individual's mental modules with the aim of achieving enhancement may be extremely harmful if the individual is considered as a closed system and their actions are abstracted from their context.[2]

For example, to what extent would enhanced altruism be an advantage for the individual? Overly empathetic persons might end up overestimating the desires and needs of others to the point of offering up themselves, their bodies and their assets (Casal, 2015). This is clearly a pathological reaction, but there is no reason why it cannot be produced by therapies aimed at enhancing empathy and altruism.

Finally, if it can be accepted, albeit tentatively, that being less selfish or more generous, more altruistic or more equitable, certainly means being morally better, then an alleged moral enhancement that would make individuals *indiscriminately* more altruistic and ipso facto morally better, and consequently make the world a better place, would entail taking a step that is not as obvious as might be thought (Baertschi, 2014). If it is true—and it is—that for the greater good it is often necessary to take actions that in other situations might be considered immoral, it is by no means certain that a general and indiscriminate moral enhancement of humanity would make the world a better place. Who would want to be governed by a president who is so empathetic as to be unable to react

[2] David DeGrazia (2014) is much more thorough and precise than his interlocutors, first of all because he sets out a fairly exhaustive list of the defects that MBE should eliminate. Secondly, he specifies where to intervene: on motives, character traits or emotions; then on cognition and understanding of what is right or the context in which one acts; and finally on behaviour making it more compliant with moral standards considered valid. Thirdly, it proposes a list of pharmacological means to be used. However, his proposal does not seem to gain legitimacy or solve the theoretical problems that undermine the very idea of moral bioenhancement.

appropriately, perhaps violently, in order to safeguard their fellow citizens from a threat? Obviously, the defence of one's own existence sometimes involves the exercise of a certain degree of violence against the aggressor, as in the case of self-defence. And again, there is little doubt that anyone who had to undergo a medical operation would want to be in the hands of an extremely effective but psychopathic surgeon rather than a sensitive and empathic surgeon, incapable of taking a calculated risk and causing a little damage to obtain a greater benefit and thus save the patient's life. Recent studies have shown that surgeons display behaviours and tendencies that verge on psychopathology, so it is easy to imagine that a therapy aimed at eliminating these "deviant" behaviours would have the guaranteed effect of reducing the number of surgeons and the skills they possess. However, as much as we can change or biomedically enhance the human body, what appears to be problematic is not so much acting on what we believe to be right, but making sure that what we consider right really is right (Casal, 2015, 1).

These clarifications aim to explain the nature of MBE and the recommended means to achieve it but, paradoxically, they make its definition even more complicated. As Casal observes, defining the administration of a hormone that increases positive feelings towards humanity as MBE would inevitably result in a similar conception of any modification producing the same effects (Casal, 2015). If moral enhancement means feeling more generous, well-disposed, empathetic and less aggressive towards humanity, then sleeping pills or alcohol would have to be regarded as MBE, something with which its supporters would surely disagree completely. Sparrow is expansive on this point, noting that without an unambiguous and shared definition of what it means to act morally, it would not be possible to avoid the annoying consequence of having to include some illegal drugs in the list of products that can legitimately be used within the MBE framework (Sparrow, 2014).

Unlike other enhancements which have the stated aim of improving individual wellbeing, moral bioenhancement performs a social function, which will be examined more closely at the end of this section. A discussion of the social aims of moral bioenhancement will make it possible to highlight the political aporia of transhumanism—the gap between its stated aims and the path enabling it to achieve them and to accurately measure its liberating potential. While other types of enhancement are left to individual assessment (in that they are partially subsidiary and verging on unnecessary), Persson and Savulescu stated that moral enhancement is a necessity stemming from the awareness that not only has scientific and technological progress failed to keep all its promises, but we have reached the edge of a pit that we have dug with our own hands by constantly producing tools and technologies that actually push us ever more violently towards extinction. Like sorcerer's apprentices unaware of the consequences of their inventions, we have created and invented extremely sophisticated techniques without understanding the risks we are exposing ourselves to, and above all without having the intellectual and moral means to utilise them in the best possible way (Persson & Savulescu, 2012a). However, it would be a mistake to believe that its supporters have converted to a mild form of technophobia on the road to Damascus. In actual fact, this criticism serves the strategy of taking a step back in order to facilitate a running jump forward: science is the problem but it is also part of the solution, it is the poison but also the antidote insofar as it provides the (pharmacological) tools that ensure it is used correctly (Fenton, 2010, 148). In reality, we are not morally suited to tackle the future that we have prepared ourselves for and it is precisely this lack of homogeneity between what we know (and know how to do) and the morality of our behaviour that puts us in jeopardy. The symptoms of the disease from which humanity is suffering are clearly traceable in the *dyscrasia* between knowledge and morality: undeniable and enormous scientific and technical advances have not been accompanied by analogous moral progress. Science has the power to eradicate virtually all the ills that afflict us but proper use of the enormous technological and scientific means developed over the centuries (i.e. without destroying every form of life on Earth) is dependent on motivations, justifications and moral concepts that have remained essentially unchanged since the dawn of time. In other words, morally we have not evolved any more than apes have, and yet we possess the powers and knowledge worthy of a *Homo sapiens* 3.0.

The moral enhancement of humanity is thus essential if it is to avoid extinction, but as Hans Jonas observed, only under a dictatorial regime can such a mandatory plan be implemented. The suspicion that MBE may lead to a reduction in individual freedoms is fuelled by scepticism of the means that can traditionally be used to enhance human morality. If there is little hope to be found in persuasion and pedagogy, in the sense of education in the rational and morally appropriate use of the enormous technological potential humanity has created, it is not because they are ineffective but because the effects of such work would be inadequate to deal with the urgent predicament of mankind. In other words, freedom and autonomy in this field would turn out to be utterly ruinous. Not only is MBE becoming an urgent imperative, but it must tackle the opposition it might provoke if it is to be achieved.

Liberal democracies are in part to blame for the problem, as their excessive liberality and tolerance is an almost insurmountable obstacle when it comes to obtaining the necessary consensus from citizens on measures that would indirectly lower their standard of living but would be indispensable in controlling access to biological or nuclear technologies potentially capable of destroying the planet and would inevitably curb individual freedoms (Persson & Savulescu, 2012a, 45). The threat posed by weapons of mass destruction cannot be reduced without greater government surveillance, although this would clearly limit the right to *privacy*, which Persson and Savulescu do not consider as such and which liberal democracies defend to the hilt (Persson & Savulescu, 2012a, 53–55). While Persson and Savulescu acknowledge that the fear of imaginary threats may produce inappropriate and disproportionate reactions, they find no reason to oppose the use of torture in extreme cases (Persson & Savulescu, 2012a, 56–58) and they give the clear impression that they would not be overly shocked even by the conversion of democracies into authoritarian regimes, the only ones with the power to eliminate or control all threats to life on Earth (Persson & Savulescu, 2012a, 60). The measures these authors advocate in order to overcome the impasse are a further source of surprise: the flip side of Persson and Savulescu's technophilia and progressivism is bleak authoritarianism and intrusive and unlimited paternalism. The first necessary measure would be a political one, designed to drastically reshape the political regimes of our societies. The recourse to drastic measures incompatible with Western democracies is thought to be fundamental in the fight against climate change: the citizens of the world's wealthiest countries would have to voluntarily reduce their standard of living, lower their consumption (eat less meat, ration water, reduce the consumption of fossil fuels, etc.), and thus, ultimately, become poorer. Under democracy, this can only be achieved through persuasion and explanation, and it is doubtful that arguments can be found to persuade the citizens of Western democracies to voluntarily lower their standard of living to ensure a better future for generations to come—or simply to make this future possible. However, the fact that the establishment of Western liberal democracies was made possible by economic growth inevitably suggests that such a strategy would directly lead to their downfall, because impoverishing them is to create conditions that are incompatible with their very existence.

From a moral standpoint, MBE would severely restrict human freedom (cf. Reichlin, 2019). If freedom implies the ability to choose between doing what should be done (an action deserving of praise) or should not be done (an action that can be blamed), it is difficult to understand how it might exist when the subject is determined to adopt a certain type of behaviour by taking a drug (Harris, 2011). External pressure to make a certain type of choice or to have a certain type of behaviour makes the moral characterisation of an action impossible, since the lack of freedom of choice between two alternatives does not allow a choice to be connoted as moral. In other words, MBE is not a moral enhancement—indeed it may not be an enhancement at all but a mechanisation of our actions in a single, predetermined (by whom?) direction.

Traditional means of ME, so to speak, make people better, not because they enable morally right actions but because they teach the proper use of freedom or, as Kant says, its critical use (where critical means a rational analysis of the limits of the use that can be made of it). The means of implementing MBE, on the other hand, teach nothing and do not facilitate the acquisition of any rational skills: herein lies the difference between a moral education that teaches living in freedom and a moral enhancement that produces specific choices. In the first case—and this is ultimately the reason why cognition and volition cannot be separated as Harris suggests—it is necessary to be able to justify the rules being enforced that are to be followed or the rules that typically are followed:

morality is dependent upon reason and freedom of choice that takes the contribution of analytical rationality into account. In the second case, chemically induced behaviour is reproduced mechanically and is neither good nor bad, fair nor unfair, because it cannot even be defined as moral.

This reasoning ultimately leads to the danger, already glimpsed in other forms of enhancement, that MBE would have even more serious implications than other forms of enhancement (cf. Pessina, 2019). What does enhancing morality by enhancing empathy actually mean? Is it an enhancement of a faculty or just a therapy to eliminate a socially dangerous defect? Hauskeller's reply, based on research showing that sociopaths have little empathy, is obvious: it is a therapy and not an enhancement (Hauskeller, 2015) that pathologises everyday normal character traits within the general context of medicalising existence (cf. Agar, 2015; Casal, 2015; Sparrow, 2014).

4.7 The Search for Immortality

All the types of enhancement seen so far are certainly important and carry weight within the transhumanist movement, but the fundamental goal of enhancement is the achievement of immortality. For all transhumanists or bioliberals, from Kurzweil, Bostrom, Sandberg and More to de Grey, and even anti-enhancement advocates like Harris, the quest for immortality is the ultimate goal that would compact and give meaning to the various lines of research on human enhancement.

Of course, the will to achieve immortality dates back to times long before the transhumanists. Ever since humanity learned to write, it has made countless efforts to overcome death, such as Bostrom's "Dragon-Tyrant" which makes human beings' lives particularly horrendous, or to complain about the fate awaiting all of humanity, to little effect in either case (Bostrom, 2005d). Practically all ancient mythologies tell the tale of a protest against the tragic mortality of human beings, starting with one of the oldest written texts that records Gilgamesh's lament over the disappearance of his friend Enkidu. In response, the goddess Siduri reminds him of the limits by which human action is inevitably confined: You will never find that life for which you are looking. When the gods created man, they allotted to him death, but life they retained in their own keeping. As for you, Gilgamesh, fill your belly with good things; day and night, night and day, dance and be merry, feast and rejoice. Let your clothes be fresh, bathe yourself in water, cherish the little child that holds your hand, and make your wife happy in your embrace; for this too is the lot of man. But Gilgamesh's despair is only one occurrence on a particularly long and universally known list of all the thoughts, texts, poems, stories and discussions occasioned by the thought of death and the way to overcome it. In the absence of the means to vanquish death, these texts share a common consolatory intent: their aim is to present the tragedy of the human condition while attempting to make it acceptable or otherwise speculate on the entirely theoretical reasons for human mortality.

Over the centuries, mythological consolation has been replaced by optimism and confidence in the powers of technology and science. In *De Dignitate* (1623) Francis Bacon names liberation from death as a goal to be achieved eventually. Sometime later, this will be considered an objective that is undoubtedly within the grasp of human potential, as emerges in the words of Mlle de Saint-Quentin published in a gazette of the time: It may be objected that we cannot live forever. Then I would gladly ask what difficulty there can be in living one year instead of another, and why, when one has lived one, one cannot live a hundred, and then a thousand, and then a hundred thousand, because it is not necessary to live the space of all this time more than that necessary to live a single hour (Mlle de Saint-Quentin, 1692–1693). Condorcet is explicit on this point: as far as the material aspect of human existence is concerned, it does not seem absurd to him "to suppose this quality of melioration in the human species as susceptible of an indefinite advancement; to suppose that a period must one day arrive when death will be nothing more than the effect either of extraordinary accidents, or of the slow and gradual decay of the vital powers; and that the duration of the middle space, of the interval between the birth of man and this decay, will itself have no assignable limit". (Condorcet, 1955, 294).

If it is true that the natural laws regulating the phenomenon of life are still unknown to us, what can prevent us from thinking that it is impossible to live for a few hundred years? Where does the idea that death is inevitable or that life must necessarily last a few dozen years come from, if not from ignorance of the laws that regulate it? However, despite the optimism of these statements, humanity had not even begun to fight death until the turn of the twentieth century, when the works of Alexis Carrel (1935), August Weismann (1892), Ilya Mečnikov (1903) and his collaborator and friend Serguei Metalnikov (1937) brought about a radical change. Previously regarded as a natural physiological process of the human body subjected to an unalterable cycle from its appearance to its disappearance, in the early twentieth century death began to be seen as a pathological state, a disease which could be healed, the effect of a morbidity that intensifies and worsens with old age, which is likewise considered a mere disease. "Death is a malfunction of the human experience," says Zoltan Istan. "It's a reversible error, a transitory cloak of emptiness, a curable disease—a highly curable disease if dealt with properly" (Istvan, 2013, 271; cf. Coeurnelle, 2013; Hayflick, 1994). In this context, immortality is not seen as a state of total invulnerability but as amortality, that is the elimination of all natural causes of biological death, while leaving unchanged the possibility that humanity may die from other causes (Mayer, 2011). The idea subsequently emerged that death is not a singular event but a biological process of cellular degeneration comprising numerous critical moments which could be tackled so as to eliminate death.

From this moment on, death becomes the natural conclusion of the accumulation of degenerative diseases that make the organism grow old and lead it to exhaustion. According to Agar, a traditional approach focused on the "diseases of ageing" was supplanted by a more ambitious one centred on "ageing as a disease" (cf. Agar, 2014, 114; Glannon, 2008b). This can be seen in the metaphor de Grey uses to explain how to strive for immortality. Just as a house that is regularly maintained and kept in good condition will better withstand all internal and external assaults, so a well-treated, properly nourished body will be able to live much longer than if it is neglected and does not receive proper attention (Kurzweil & Grossman, 2004a, 3). Although this is an apparently reasonable metaphor, the idea that death is the end result of ageing (understood as an accumulation of diseases and therefore a pathological process in itself) is far from being accepted by the scientific community. Hayflick proposes criteria for distinguishing ageing from disease in general and from diseases that arise in this stage of existence. Unlike the latter, ageing affects all individuals indiscriminately, it begins only after the end of the reproductive period and it is present in all "domesticated" individuals. In short, it is an undesirable effect of civilisation which has improved living conditions and above all eliminated external dangers and threats such as predators, intra-social competition and atmospheric conditions and thus enabled organisms to live much longer than if they were left in their natural state. According to Hayflick, age-related diseases and the ageing process are two completely different things: while the former should be treated like any other disease, the latter is a natural, non-pathological process (Barazzetti, 2011, 337); while ageing is not a disease, it is clear that there are diseases related to ageing.

However, therapies for such pathologies are important for research into immortality because they constitute one of the three scenarios on which life-extension research is focused: together with research on the "compression of morbidity", the aim is to identify the means to achieve "decelerated ageing", or even "arrested ageing" (Post & Binstock, 2004). The compression of morbidity is explicitly focused on ensuring a healthier and thus qualitatively better life span which might also lead to an extension of life, although researchers predict that it could never exceed the maximum life span observed so far and, in any case, would not significantly extend it or eliminate death. A more ambitious project aims to slow down ageing, which could significantly increase the human life span beyond its current limits. Finally, the transhumanist scenario sees research focused on the causes of ageing in an attempt to arrest it or even eliminate it by allowing "the achievement of a perpetually youthful physiological state". Human beings who no longer age would find themselves in a state of "virtual immortality", and their existence would be threatened only by external causes (Barazzetti, 2011, 338–339). De Grey makes an even more ambitious proposal that it should be possible to eliminate or significantly reduce the causes of involuntary death (De Grey, 2004a, 30).

Although immortality is the subject of a considerable scientific research effort, results have been quite scarce so far (cf. De Grey &

Rae, 2007), although it cannot be said that supporters of the imminent achievement of immortality lack ideas or funding. So important is this goal that research on the causes of old age and death is intensifying, given the evident lack of tangible results, and it seems likely that the near (though not immediate) future will yield therapies and practices to ensure a life span far in excess of the current record of 125 years or achieve the dream that humanity has always pursued: immortality. This has been the long-standing objective of Kurzweil, de Grey and the others who are trying to survive as long as possible through somewhat rudimentary remedies of extremely limited efficacy, starting with hibernation (which is not so much a way to combat death but a survival technique until suitable therapies are found, as is indicated in R. Ettinger's classic *The Prospect of Immortality* (1962) (cf. Hughes, 2001). Their hope is to see the development of more effective anti-ageing therapies, which should make it possible to further extend life expectancy and live on until an immortality therapy is finally discovered.

According to Kurzweil, technological progress is advancing at such a rapid pace that the means to ensure immortality is expected to arrive in the foreseeable future. The important thing is to be able to see this moment. Kurzweil's "Longevity Programme" aims to do just that in three stages, starting with "present-day therapies and guidance that will enable you to remain healthy long enough to take full advantage of the construction of the Second Bridge". Kurzweil indicates a number of methods for extending human life, including calorie restriction, growth hormones, the insulin-like growth factor (IGF-I), DHEA, melatonin, testosterone, progesterone and oestrogens, interventions in oxidation processes, interventions in telomere activation, genetic manipulation and stem cell research, lots of exercise and prevention (Rose & alii, 2004), or as summarised by the acronym TRANSCEND, the title of the book that encapsulates Kurzweil's methodology: "Talk with your doctor, Relaxation, Assessment, Nutrition, Supplements, Calorie reduction and weight loss, Exercise, New Technologies, Detoxification" (Kurzweil & Grossman, 2004b). Following this programme should make it possible to reach the second level, that is the biotechnological revolution that will reveal all the secrets of cell functioning, and subsequently the third level, that is the "nanotechnology-AI (artificial intelligence) revolution" that will allow us to "rebuild our bodies and brains at the molecular level" (Kurzweil & Grossman, 2004a, 15).

In this third phase, nanotechnologies guided by artificial intelligence will enable full control of all human biological processes so as to avoid disease, ageing and death. It is this reasoning (or this hope) that underpins the "longevity escape velocity" coined by de Grey, which states that the pace of biomedical progress extending human life every day is such that the first human being who will enjoy immortality has probably already been born (De Grey, 2004a).

Unlike Kurzweil, de Grey provides elaborate scientific details. He claims there are only a few differences ("seven deadly things", which can mostly be eliminated) between the cells of a young person and those of an elderly person: these consist of a decline in the number of cells in certain tissues and an accumulation of unwanted cells of certain types, of mutations in our chromosomes, of mutations in our mitochondria, of random cross-links between long-lived extracellular proteins and of chemically inert but bulky "junk" in our lysosomes and in extracellular spaces (De Grey, 2004a, 33, 2007, 43; cf. Agar, 2010, 88–89). However, regardless of the identification of the different biological processes leading to old age and death, the basic structure of transhumanist arguments remains the same: take control of biological processes to guide and direct them through the use of technology (Dickel & Frewer, 2014, 124).

Although this research is undeniably important and may be an epochal turning point in the history of mankind, surprisingly support is not unanimous. Bostrom and Ord (2006) observe that innovation is always resisted precisely because it is innovative and regardless of any potential benefits (Agar, 2010, 133–140) which, in the case of immortality, would be incalculable.

In the *Fable of the Dragon-Tyrant*, in which the dragon symbolises death, Bostrom summarises not only the advantages for humanity that would result from the "death of death" but also the resistance that this research encounters in the scientific world and in some areas of public opinion. This resistance is totally incomprehensible, considering that immortality would be not so much the acquisition of an indefinite life span but a new birth: humanity would return to a child-like condition from which to restart and build a better world, as announced by the king who defeated the dragon-tyrant in Bostrom's eponymous fable (Bostrom, 2005c). Those in favour see immortality as the final victory over absolute evil and find it hard to understand how anyone can be opposed to such a quest, especially as immortality has always been considered not simply *a* good, but *the* good to be achieved at all costs. Those who are against it fear that the search for this absolute good will result in an equally absolute evil that, in the worst case scenario, could bring about the extinction of the human race (Post & Binstock, 2004). On the other hand, as Bauman recalls, Elias Canetti asked a question that never ceases to resound as a warning, "How many people will discover that life is worth living once they no longer have to die?", as he believes the answer is very few (Bauman, 1992, 6). In actual fact, there does not seem to be a connection between the fear of dying and the will to be immortal, as Hauskeller argues (2013, 89–92).

Apart from individual reactions to what each person wants for themselves, objections have often been raised about the individual and social impact that immortality could have.

From the individual standpoint there are, first of all, fears of repercussions on personal identity: as an individual would live several lives and encounter ever greater difficulties in reconstituting their identity in a single narrative, they might end up being overwhelmed by the number of existences lived. The consequence is that it would no longer be "me" that is immortal, which would effectively nullify my existence (Glannon, 2002). This argument has the disadvantage of being based on a specific conception of personal identity, founded on Parfit's psychological and functional continuity, which is not universally accepted. Alternative theories of personal identity that seem perfectly compatible with immortality have been proposed by such authors as Olson, who claims "we are basically human animals" (Barazzetti, 2011, 342–343; DeGrazia, 2005; Olson, 1997; cf. Agar, 2014, ch. 4; Schechtman, 2014).

Second, as many writers have envisioned, it is not necessarily true that an infinite life would be desirable: it might be extremely boring and, ultimately, useless. Bernard Williams exemplifies this view by citing Janacek's *The Makropoulos Case*, whose protagonist lives for several hundred years before coming to the conclusion that her infinite life has become cold, boring and, in the final analysis, meaningless. The infinite repetition of the same things has made them pointless and senseless, so she decides to die (Williams, 1976, 82; cf. Adams, 1979; Agar, 2010, 112; Gems, 2003; Hauskeller, 2013, 89–113; Jonas, 1996, 87–98).

Third, the notion of the "naturalness of ageing" is raised, that is the idea that ageing has the specific function in evolution of eliminating individuals who cannot reproduce leaving room for fertile young people (Barazzetti, 2011, 343).

Finally, Agar fears that the people he defines as "negligibly senescent" could end up withdrawing from the world and rejecting its risks (Agar, 2010, 123), which would have a major impact on the world and also on scientific research, which might diminish and even come to a halt.

The general objection to these arguments is that the negative psychological consequences resulting from immortality are such because we are not immortal. It is likely that "our natures and psychologies would surely change if we lived forever" (Temkin, 2011, 350), and in any case, we cannot begin to imagine the psychology of a posthuman being who has undergone changes that make them completely different from the beings we currently are.

From the social point of view, it has been suggested that immortality might contribute to social injustice: as life extension is expensive, it will not be within everyone's reach (cf. Barazzetti, 2011, 343–344; Harris, 2007, 59–71).

The economic problem would stem from the increased costs incurred by having to maintain a healthy population continuously engaged in a fight against ageing. There would also be inevitable problems of social justice caused by the existence of two populations, one immortal and the other unable to afford the treatment and consequently condemned to die, but also by the overpopulation of the planet and the scarcity of natural resources.

Despite these objections, there is no one who does not see the interest in continuing research aimed at eliminating death (Overall, 2003). In order to overcome the resistance of the most reluctant critics, the advocates of immortality have tapped their imagination and boldly described the paradisiacal life of humanity freed from the terrible scourge of death, forgetting however, among other things, Jonathan Swift's description of immortal life, although, in defence of the "immortalists", this is certainly not the immortality they seek. During his stay with the inhabitants of Luggnagg, Gulliver learns of the existence of the struldbrugs, beings who, quite by chance, are immortal. When he is asked what he thinks the life of these individuals is like, Gulliver imagines rich, wise people dedicated to the study and education of humanity with a view to its material and moral improvement. What he does not know is that the struldbrugs do grow old like everyone else, but this process lasts for eternity. Ageing brings with it the whole panoply of ills that afflict all individuals, such as baldness, tooth loss, diminished strength, reduced memory capacity and weakened intellectual faculties, with the result that they can no longer read and cannot reach the end of a sentence without forgetting the beginning (Jonas, 1962).

It must be stressed, however, that the immortality that transhumanists strive to achieve is completely different. It is not a question of growing old indefinitely, as in the myth of Tithonus, but of arresting the growth process before it begins to decline. The immortalists undoubtedly relate more to the much simpler, not to say simplistic, situation envisioned by Nick Bostrom. Freed from the dragon-tyrant that feeds on men and living in a perfect state of health in a *potentially* infinite life, humanity will have finally attained a kind of heaven on Earth, where every day is full of joy and pleasure. At this point, music, art, games, language, relationships and, basically, all human practices will reach such a level of perfection as to produce a feeling of wellbeing and happiness that is impossible for us to describe. According to Bostrom, engaging in these new forms of activity will reveal "how every fiber of your body and mind is stretched to its limit in the most creative and imaginative way, and you are creating new realms of abstract and concrete beauty that humans could never (concretely) dream of". Furthermore, compassion for all forms of life will be brought to a level such that it will be a pleasure to engage in any activity that can reduce the "suffering of animals in their natural environment", which will be respected and protected. In short, although not everything is perfect yet ("Things are getting better", writes Bostrom (2009a, 111), "but already each day is fantastic".

De Grey also tries to imagine what the life of an immortal being or an individual who need not suffer an involuntary death might be like. For the moment, when you retire, you retire forever. We're sorry for old people because they're going downhill. There will be no real moral or sociological requirement to do that. Sure, there is going to be a need for Social Security as a safety net just as there is now. But retirement will be a periodic thing. You'll be a journalist for 40 years or whatever and then you'll be sick of it and you'll retire on your savings or on a state pension, depending on what the system is. So, after 20 years, golf will have lost its novelty value, and you'll want to do something else with your life. You'll get more retraining and education, and go and be a rock star for 40 years, and then retire again and so on (Agar, 2010, 108).

Mutatis mutandis, these visions do not seem to be substantially different from the reality and the pleasures that typify, according to the goddess Siduri, the happy life of a mortal being like Gilgamesh. However, to some they may also seem empty and ultimately useless. If criticising the search for immortality, which finally seems to be within reach, is a rhetorical exercise, whose supporters' sincerity or adherence cannot be evaluated, it is also true that the depiction of immortal life, whose all too rosy colours call to mind a Hollywood blockbuster, is rather disappointing. The blame could be placed on a certain lack of imagination on the part of the transhumanists who, too busy pursuing their goals, lose sight of the reason for so much expenditure of energy and means. But in actual fact it is more likely that these rather disappointing descriptions of immortal life are a symptom of transhumanism's belonging to modernity, which incidentally might well make it much less revolutionary than we think. This view is based on a suggestion made by Arendt, who observes that, for the first time, humanity has the real power to detach itself and abandon the human context and human activities, and he imagines that it can colonise space or free itself from the slavery of work. He notes that this society, which is laboriously freeing itself from work, is a society of workers, that is a society that "does no longer know of those other higher and more meaningful activities for the sake of which this freedom would deserve to be won" (Arendt, 1998, 5). The suspicion hanging over the search for immortality, which in and of itself is perhaps not totally unjustified, is that humanity, although immortal, will not have gained any happiness; indeed, it will have worsened its situation in that it will now be condemned to live for millennia without knowing why.

With regard to the kind of human that an immortal being would be, opinions differ, but they all basically agree that in one way or another it would be an individual whose biological elements are hybridised with different kinds of technology, running physiological processes. Stanley Shostak, who is not a transhumanist, depicts a rather futuristic and probably unachievable being that certainly contains elements consistent with transhumanist theories. Having reviewed all the scientific research on the reasons for ageing and the biological causes of death and how to eliminate them, Shostak states that, as immortality could be achieved by arresting growth, the best time to stop it would be prepuberty because of its greater dynamic balance (Shostak, 2002, 163). This arrested growth could be easily achieved by inserting a new organ ("a generator of self-renewing stem-cells derived from a clone") into the body (Shostak, 2002, 166). The consequence would be a population—or at least the part of the population that can afford this new organ—made up of immortal children, adolescent cyborgs, beings that would never complete their growth but would remain in a state of sexual immaturity. They would thus be unable to reproduce but would basically be in a state of life in which everything is about to come true, and therefore happily stuck in a situation of pure potential (Shostak, 2002, 164). That such an organ can be invented, that it can be implanted, that it actually works and that immortality can therefore be achieved in this way is, ultimately, of little importance. What is confirmed is the idea that the human beings of the future will increasingly be cyborgs, "humans" hybridised with technologies aimed at enhancing their potential. Shostak seems to be the spokesperson of the widely shared desire to live in eternal youth, which has recently turned into the desire to remain immature, as represented by the character of Peter Pan.

However, are the various immortalists like Bostrom, de Grey, Shostak and Harrington (1969) right in thinking of immortal life as a simple infinite extension of human life or do they completely misrepresent the consequences of immortality? Their descriptions of immortal lives are rather flat and reassuring and do not take into account the complexity of human life or humanity's incessant effort to make sense of the senseless, which is actually clustered in all human constructs.[3] Moreover, just as it is impossible to understand the psychology of immortal beings, it should be just as impossible to describe the kind of existence that immortals might lead and to imagine not only the kind of society they would live in but whether they would live in a society at all. Is it realistic to think that the elimination of death would lead to an existence like the one described by Bostrom, Kurzweil and de Grey? Perhaps. But it is equally probable that the supporters of immortality, whether deliberately, superficially or carelessly, underestimate the role that death has played in the construction of human life and therefore the changes that its elimination would entail. There is a widely shared view that the fight against death has produced all human artefacts, whether these be philosophy, culture or the political, social and economic structure of our societies; if this is true, what would remain if death and the awareness of mortality were removed.[4] What would the human world be like? Is it possible that immortality might simply produce an exponential multiplication of the very elements of a finite life?

Shostak's considerations suggest that time is a succession of present moments for immortals, that the arrow of time has been abolished from its own extension to infinity, and that time flows uniformly like an eternal present. Shostak does not go so far as to describe the daily life of these eternal youngsters but presents some positive and negative indications on this particularly significant aspect of immortality achieved through the abolition of time (Shostak, 2002, 205). While it is clear that time would become infinitely accessible, the consequences are far from being completely positive. The present would totally absorb the future

---

[3] In a work succinctly and emblematically entitled *The Temptation to Exist*, Émile Cioran highlights the existential aspect of death: "Life, far from being what Bichat once called an ensemble of functions for resisting death, is rather an ensemble of functions for bearing us toward it. Our substance diminishes with every step, yet it is of this very diminution that all our efforts should tend to make a stimulant, a principle of efficacity. Those who cannot benefit from their possibilities of nonexistence are strangers to themselves: puppets, objects 'furnished' with a self, numbed by a neutral time that is neither duration nor eternity" (Cioran, 1968, 113).

[4] For example: "Death is the real inspiring genius or Musagetes of philosophy, indeed, without death there would hardly have been any philosophising" (Schopenhauer, 1969, 1021; cf. Bauman, 1992).

and the past, making the passage from one into the other utterly meaningless: in other words, and keeping Heidegger's lesson in mind, if being is time, then where there is no more time there would be no more being. Being would be completely and definitively devoid of meaning and this would have potentially disastrous consequences, as will be shown below in an attempt to understand what such an existence might be like.

An initial and partial indication is given by Rousseau in his essay on the *Origin of Inequality*, which describes the "savages" who, although not biologically immortal, nevertheless live in a sort of eternal present because they have no cognition of death and have, therefore, acquired a form of immortality. Rousseau's savages are in every way similar to animals and, like them, are subject to physical needs and to the search for the only goods they know: "food, a female and rest", while avoiding the only evils they fear, "pain and hunger". Rousseau firmly adds: "I say pain, and not death; for an animal will never know what it is to die, and the knowledge of death, and its terrors, is one of man's first acquisitions on moving away from the animal condition" (Rousseau, 1755, 142). In short, Rousseau believes that humanity, deprived of the knowledge of death, is incapable of history, culture, politics and science, he is nothing more than an animal and, like them, he is immortal because he is unaware of his own mortality. Moreover, leaving the state of nature can only be achieved if one is aware that one's own existence is continually threatened by others—without being able to say which is the cause and which is the effect: for Hobbes the covenant aims specifically to eliminate the uncertainties of a life subjected to frequent threats of imminent death. Whereas Hegel claims that if there is no possibility that a man might risk his life in a mortal struggle with another, he could not begin the dialectic of recognition, that is the reason why a man chooses to live in society.

But if this is the case, then immortality might not result in the beautiful inventions of Bostrom and Shostak, the bold and enlightened theories of Swift, Bacon and Condorcet, or to the nihilism of Capek and Adams, but would instead produce people totally unaware of death who would live like Rousseau's savages. The world and the life of these immortals would be much more similar to those described by Borges, who sees them as troglodytes: completely lost "in thought, in pure speculation", "they hardly perceived the physical world". Knowing that an immortal has the time to live every experience without dying, they "had attained the perfection of tolerance and almost that of indifference", because in the perspective of an infinite world everything appears identical. In this world "there are no moral or intellectual merits": if in ours, Homer achieved a great feat in composing the *Odyssey*, "with infinite circumstances and changes, the impossible thing is not to compose the *Odyssey*, at least once" (Borges, 1957, 117). While for us, all acts have "the value of the irretrievable and the perilous", "among the Immortals, on the other hand, every act (and every thought) is the echo of others that preceded it in the past, with no visible beginning, or the faithful presage of others that in the future will repeat it to a vertiginous degree" (Borges, 1957, 118). Following in the footsteps of Borges would at least suggest the risk that a society of immortals might be far from the daydreams hatched by the imagination of a Bostrom or a Shostak. If our society is a society of mortals, not to say the dying, who in their finite condition find the impulse to build a comfortable bed suitable for their finitude in line with the advice given to Gilgamesh, the end of death will also result in the end of our world and the disappearance of all the little pleasures that we allow ourselves while waiting for the inevitable.

References

Adams, D. (1979). *The Hitchhiker's Guide to the Galaxy*. Pan Books.

Agar, N. (2010). *Humanity's End*. MIT Press.

Agar, N. (2014). *Truly human enhancement: A philosophical defense of limits*. MIT Press.

Agar, N. (2015). Moral bioenhancement is dangerous. *Journal of Medical Ethics, 41*(4), 343–345.

Alexandre, L. (2011). *La mort de la mort*. Lattès.

Arendt, H. (1998). *The human condition* (2nd ed.). University of Chicago Press.

Baertschi, B. (2009). Devenir un être humain accompli: idéal ou cauchemar. In J.-N. Missa & L. Perbal (Eds.), *«Enhancement». Ethique et philosophie de la médecine d'amélioration* (pp. 79–95). Vrin.

Baertschi, B. (2014). Neuromodulation in the Service of Moral Enhancement. *Brain Topography, 27*(1), 63–71.

Barazzetti, G. (2011). Looking for the fountain of youth: Scientific, ethical, and social issues in the extension of human lifespan. In J. Savulescu, R. ter Meulen, & G. Kahane (Eds.), *Enhancing human capacity* (pp. 335–349). Wiley-Blackwell.

Bauman, Z. (1992). *Mortality, immortality and other life strategies.* Polity Press.

Baumgartner, T., Heinrichs, M., Vonlanthen, A., Fischbacher, U., & Fehr, E. (2008). Oxytocin shapes the neural circuitry of trust and trust adaptation in humans. *Neuron, 58*(49), 639–650.

Beauchamp, T. L. (2015). Are we unfit for the future? *Journal of Medical Ethics, 41*(4), 346–348.

Berghmans, R., ter Meulen, R., Malizia, A., & Vos, R. (2011). Scientific, ethical, and social issues in mood enhancement. In J. Savulescu, R. ter Meulen, & G. Kahane (Eds.), *Enhancing human capacity* (pp. 153–165). Wiley-Blackwell.

Berlin, I. (2002). *Liberty*. Oxford University Press.

Borges, J.-L. (1957). *The Immortal*. In Borges J.-L. 1962. *Labyrinths and other stories*. New Directions.

Bostrom, N. (2005b). In defense of posthuman dignity. *Bioethics, 19*, 202–214.

Bostrom, N. (2005c). Transhumanist values. *Review of Contemporary Philosophy, 4*, 87–101.

Bostrom, N. (2005d). The fable of the dragon tyrant. *Journal of Medical Ethics, 31*, 273–277.

Bostrom, N. (2008a). Letter from Utopia. *Journal of Evolution and Technology, 19*(1), 67–72.

Bostrom, N. (2009a). Why I want to be a posthuman when I grow up. In A. M. Cutter, B. Gordijn, G. E. Marchant, A. Pompidou, & R. Chadwick (Eds.), *Medical enhancement and posthumanity* (pp. 107–136). Springer.

Bostrom, N., & Ord, T. (2006). The reversal test: Eliminating status quo bias in applied ethics. *Ethics, 116*(4), 656–679.

Bostrom, N., & Savulescu, J. (2009). Human enhancement ethics: The state of the debate. In J. Savulescu & N. Bostrom (Eds.), *Human enhancement* (pp. 1–22). Oxford University Press.

Buchanan, A. (2011b). *Beyond humanity?* Oxford University Press.

Cabrera, L. Y. (2015). *Rethinking human enhancement: Social enhancement and emergent technologies*. Palgrave Macmillan.
Carrel, A. (1935). *L'homme, cet inconnu*. Plon.
Carter, J. A., & Gordon, E. C. (2015). On cognitive and moral enhancement: A reply to Savulescu and Persson. *Bioethics, 3*(29), 153–161.
Casal, P. (2015). On not taking men as they are: Reflections on moral bioenhancement. *Journal of Medical Ethics, 41*(4), 340–342.
Churchland, P. (2012). *Braintrust: What neuroscience tells us about morality*. Princeton University Press.
Cioran, E. (1968). *The temptation to exist*. Arcade.
Coeckelbergh, M. (2013a). *Human Being @ Risk. Enhancement, technology, and the evaluation of vulnerability transformations*. Springer.
Coeurnelle, D. (2013). *Et si on arrêtait de vieillir!* Fyp éditions.
Critical Art Ensemble. (1998). *Flesh Machine: Cyborgs, Designer Babies, and New Eugenic Consciousness*. Autonomedia.
Damasio, A. (1994). *Descartes' error: Emotion, reason, and the human brain*. Putnam Publishing.
de Condorcet, M. (1955). *Sketch for a historical picture of the progress of the human mind*. Weidenfeld & Nicolson.
De Grey, A. (2004a). Escape velocity: Why the prospect of extreme human life extension matters now. *PLOS Biology, 2*(6), 0723–0726.
De Grey, A., & Rae, M. (2007). *Ending aging*. Saint Martin's Press.
DeGrazia, D. (2005). Enhancement technologies and human identity. *Journal of Medicine and Philosophy, 30*(3), 261–283.
DeGrazia, D. (2014). Moral enhancement, freedom, and what we (should) value in moral behaviour. *Journal on Medical Ethics, 40*(6), 361–368.
Delfin, D. (2019). An artist's creative process: A model for conscious evolution. In N. Lee (Ed.), *The transhumanism handbook* (pp. 587–602). Springer.
Devereaux, M. (2008). Cosmetic surgery. In B. Gordijn & R. Chadwick (Eds.), *Medical enhancement and posthumanity* (pp. 169–174). Springer.
Dickel, S., & Frewer, A. (2014). Life extension: Eternal debates on immortality. In R. Ranisch & S. L. Sorgner (Eds.), *Post- and transhumanism: An introduction* (pp. 119–131). Peter Lang.
Douglas, T. (2008). Moral enhancement. *Journal of Applied Philosophy, 25*(3), 228–245.
Douglas, T. (2013). Moral enhancement via direct emotion modulation: A reply to John Harris. *Bioethics, 27*(3), 160–168.
Elliott, C. (1999). *A philosophical disease*. Routledge.

Elliott, C. (2004). *Better than well: American medicine meets the American dream.* Norton & Company.
Engelhardt, H. T. (1991). *Bioethics and secular humanism: The search for a common morality.* SCM Press-Trinity Press International.
Fenton, E. (2010). The perils of failing to enhance: A response to Persson and Savulescu. *Journal of Medical Ethics, 36*(3), 148–151.
Freedman, C. (1998). Aspirin for the mind? Some ethical worries about psychopharmacology. In E. Parens (Ed.), *Enhancing human traits: Ethical and social implications* (pp. 135–150). Georgetown University Press.
Gems, D. (2003). Is more life always better? The new biology of aging and the meaning of life. *Hastings Center Report, 4*(33), 31–39.
Glannon, W. (2002). Identity, prudential concern, and extended lives. *Bioethics, 3*(16), 266–283.
Glannon, W. (2008b). Decelerating and arresting human aging. In B. Gordijn & R. Chadwick (Eds.), *Medical enhancement and posthumanity* (pp. 175–190). Springer.
Glover, J. (1988). *I: The philosophy and psychology of personal identity*. Penguin.
Gordijn, B. (2006). Converging NBIC technologies for improving human performance: A critical assessment of the novelty and the prospects of the project. *The Journal of Law, Medicine & Ethics, 34*(4), 726–732.
Harrington, A. (1969). *The immortalist*. Random House.
Harris, J. (2007). *Enhancing evolution.* Princeton University Press.
Harris, J. (2011). Moral enhancement and freedom. *Bioethics, 25*(2), 102–111.
Harris, J. (2013). Ethics is for bad guys! Putting the 'moral' into moral enhancement. *Bioethics, 27*(3), 169–173.
Hayflick, L. (1994). *How and why we age.* Ballantine Books.
Hauskeller, M. (2013). *Better humans: Understanding the enhancement project*. Acumen.
Hauskeller, M. (2015). Being good enough to prevent the worst. *Journal of Medical Ethics, 41*(4), 289–290.
Hauskeller, M. (2016). *Mythologies of transhumanism.* Palgrave Macmillan.
Herath, D., Kroos, C., & Stelarc. (2016). *Robots and art: Exploring an unlikely symbiosis.* Springer.
Hildt, E., & Franke, A. (2013). *Cognitive enhancement: An interdisciplinary perspective.* Springer.
Housden, C. R., Morein-Zamir, S., & Sahakian, B. J. (2011). Cognitive enhancing drugs: Neuroscience and society. In J. Savulescu, R. ter Meulen, & G. Kahane (Eds.), *Enhancing human capacity* (pp. 113–126). Wiley-Blackwell.

Hughes, J. (2001). The future of death: Cryonics and the telos of liberal individualism. *Journal of Evolution and Technology, 6*(1), 1–30.
Istvan, Z. (2013). *The transhumanist wager*. Futurity Imagine Media.
Jonas, H. (1962/2001). Immortality and the modern temper. In *The Phenomenon of life* (pp. 262–281). Northwestern University Press.
Jonas, H. (1996). The burden and blessing of mortality. In *Mortality and morality: A search for the good after Auschwitz* (pp. 87–98). Northwestern University Press.
Jones, G., & Whitaker, M. (2012). Transforming the human body. In C. Blake, C. Molloy & S. Shakespeare (Eds.), *Beyond human: From animality to transhumanism* (pp. 254–279). Bloomsbury.
Kurzweil, R., & Grossman, T. (2004a). *Fantastic voyage: Live long enough to live forever*. Rodale Books.
Kurzweil, R., & Grossman, T. (2004b). *Transcend: Nine steps to living well forever*. Rodale Books.
Lavazza, A., & Reichlin, M. (2019). Moral enhancement. *Topoi, 38*(1).
Leibniz, G. W. (2005). *Confessio Philosophi. Papers concerning the problem of evil, 1671–1678*. Yale University Press.
Lewis, S. C. (1943). *The abolition of man*. Samizdat University Press.
Mayer, C. (2011). *Amortality: The pleasures and perils of living agelessly*. Random House.
Mečnikov, I. (1903). *Études sur la nature humaine. Essais de philosophie optimiste*. Masson.
Merleau-Ponty, M. (2014). *Phenomenology of perception*. Routledge.
Metalnikov, S. (1937). *La lutte contre la mort*. Gallimard.
Miah, A. (2011). Physical enhancement: The state of the art. In J. Savulescu, R. ter Meulen, & G. Kahane (Eds.), *Enhancing human capacity* (pp. 266–273). Wiley-Blackwell.
Mlle de Saint-Quentin. (1692–1693). Traité pour établir la possibilité de l'immortalité corporelle. *Mercure Galant, 33–34*, 120–124.
More, M. (1996). *Transhumanism: Towards a futurist philosophy*. https://web.archive.org/web/20051029125153/http://www.maxmore.com/transhum.htm. Last accessed 26 April 2020.
Niyadurupola, G. (2008). Better brains. In L. Zonneveld (Ed.), *Reshaping the human condition: Exploring human enhancement* (pp. 77–88). Rathenau Institute.
Nussbaum, M. C. (2001). *Upheavals of thought: The intelligence of emotions*. Cambridge University Press.

Olson, E. T. (1997). *The human animal: Personal identity without psychology*. Oxford University Press.
Overall, C. (2003). *Aging, death and human longevity*. University of California Press.
Parens, E. (1998a). *Enhancing human traits: Ethical and social implications*. Georgetown University Press.
Parens, E. (1998b). Is better always good? The enhancement project. In E. Parens (Ed.), *Enhancing human traits: Ethical and social implications* (pp. 1–28). Georgetown University Press.
Persson, I., & Savulescu, J. (2012a). *Unfit for the future: The need for moral enhancement*. Oxford University Press.
Pessina, A. (2016). *L'Io insoddisfatto. Tra Prometeo e Dio*. Vita & Pensiero.
Pessina, A. (2019). A clay person: The promises of moral bioenhancement. In A. Lavazza & M. Reichlin (Eds.), Moral enhancement. *Topoi, 38*(1), 87–93.
Post, S. G., & Binstock, R. H. (2004). *The fountain of youth*. Oxford University Press.
Rakić, V. (2014). Voluntary moral enhancement and the survival-at-any-cost bias. *Journal of Medical Ethics, 40*, 246–250.
Reichlin, M. (2019). The moral agency argument against moral bioenhancement. In A. Lavazza & M. Reichlin (Eds.), Moral enhancement. *Topoi, 38*(1), 53–62.
Rose, N. (2007). *The politics of life itself: Biomedicine, Power, and subjectivity in the twenty-first century*. Princeton University Press.
Rose, S. (2006). Brain Gain. In P. Miller & J. Wilsdon (Eds.), *Better humans? The politics of human enhancement and life extension* (pp. 69–78). DEMOS.
Rose, M. R., & alii. (2004). *The scientific conquest of death: Essays on infinite lifespans*. Libros en Red.
Rothman, S., & Rothman, D. (2003). *The pursuit of perfection: The promise and perils of medical enhancement*. Pantheon Books.
Rousseau, J.-J. (1755/1997). *Discours on the origin and foundation of inequality among men or second discourse*. In *The discourse and other early political writings* (pp. 111–222). Cambridge University Press.
Sandberg, A. (2003). Morphological freedom: Why we not just want it, but need it. In M. More & N. Vita-More (Eds.), *The transhumanist reader* (pp. 56–64). Wiley Blackwell.
Sandberg, A. (2011). Cognition enhancement: Upgrading the brain. In J. Savulescu, R. ter Meulen, & G. Kahane (Eds.), *Enhancing human capacity* (pp. 71–91). Wiley-Blackwell.

Sandberg, A., & Savulescu, J. (2011). The social and economic impacts of cognitive enhancement. In J. Savulescu, R. ter Meulen, & G. Kahane (Eds.), *Enhancing human capacity* (pp. 92–112). Wiley-Blackwell.

Savulescu, J., Sandberg, A., & Kahane, G. (2011). Well-Being and Enhancement. In J. Savulescu, R. ter Meulen, & G. Kahane (Eds.), *Enhancing human capacities* (pp. 3–18). Wiley-Blackwell.

Schechtman, M. (2014). *Staying Alive: Personal identity, practical concerns, and the unity of a life*. Oxford University Press.

Schopenhauer, A. (1969). *The world as will and representation*. Dover.

Shostak, S. (2002). *Becoming immortal*. State University of New York Press.

Sparrow, R. (2014). Better living through chemistry? A reply to Savulescu and Persson on 'Moral Enhancement'. *Journal of Applied Philosophy, 31*(4), 23–32.

Temkin, L. (2011). Is living longer living better? In J. Savulescu, R. ter Meulen, & G. Kahane (Eds.), *Enhancing human capacity* (pp. 350–367). Wiley-Blackwell.

ter Meulen, R. (2015). The moral ambiguity of human enhancement. In S. Bateman, J. Gayon, S. Allouche, J. Goffette, & M. Marzano (Eds.), *Inquiring into human enhancement: Interdisciplinary and International Perspectives* (pp. 86–99). Palgrave Macmillan.

Tirosh-Samuelson, H. (2011b). Engaging transhumanism. In G. R. Hansell & W. Grassie (Eds.), *H±: Transhumanism and its critics* (pp. 10–20). Metanexus Institute.

Turner D. & Sahakian B. 2008. The Cognition-enhanced Classroom. In *Reshaping the Human Condition: Exploring Human Enhancement*, eds. Zonneveld L. & alii, 107–114. The Hague: Rathenau Institute.

Vita-More, N. (2013). Aesthetics: Bringing the arts & design into the discussion of transhumanism. In M. More & N. Vita-More (Eds.), *The transhumanist reader* (pp. 18–27). Wiley Blackwell.

Weismann, A. (1892). *Essais sur l'hérédité et la sélection naturelle*. Reinwald.

Williams, B. (1976). The Markopoulos case: Reflections on the tedium of immortality. In B. Williams (Ed.), *Problems of the self* (pp. 82–100). Cambridge University Press.

5

Problems of Transhumanism

5.1 Criticism of Enhancement

The loudly proclaimed positive consequences of enhancement do not seem sufficient to justify its moral legitimacy. On the contrary, some philosophers and theorists have focused almost exclusively on the dangers and negative potential they entail. Early targets of such virulent criticism and even declarations of revulsion were genetic manipulation/selection and, above all, attempts at cloning, although these are not central to transhumanism. Leon Kass not only exploited the negative reactions to depictions of dystopia, such as Huxley's *A Brave New World* which in his opinion foreshadows the developments of transhumanism, but also highlighted how these practices, first and foremost cloning, are contrary to common sense and provoke general reactions of disgust and repugnance. However, a critique based on emotional reactions sheds doubt on the strength of its arguments: this can be easily countered by pointing out, as Kass himself acknowledges, that many practices which initially aroused the same negative feelings are now part of the cultural landscape (Giubilini & Sanyal, 2016, 7). Kass's position lacks the fundamental rigour needed to justify it and thus remains a mere emotional reaction

F. P. Adorno, *The Transhumanist Movement*,
https://doi.org/10.1007/978-3-030-82423-5_5

with no greater value than any other position (cf. Giubilini & Sanyal, 2016, 8; Roache & Clarke, 2009).

In addition to these emotional reactions, there has been further criticism of the human enhancement techniques advocated by transhumanists that has targeted several assumptions and some negative effects that might result from their use. On these occasions, doubts were raised over the safety of such techniques (cf. Haisma, 2011) and it was suggested that they might compromise the concepts and behaviour that underpin moral codes generally and would thus impact distributive justice and social equity (cf. Levy, 2007; Tamburrini & Tännsjö, 2011). Many such criticisms target specific enhancement practices (discussed in the following chapter) while others concern enhancement as such, which will be discussed here. The focus will be on objections to human enhancement practices that endanger human nature either by breaking the limits of that nature or by removing any dignity that it could be attributed with, which is the premise on which humanity's moral and political concepts are largely based.

First of all, it has been objected that those who employ NBICs for purposes of human enhancement with the specific aim of shaping matter to their own will are behaving like the sorcerer's apprentice or, more dramatically, manifesting the will to "play God" or, in its secularised version, to go so far as to intervene in natural processes that cannot yet be adequately understood (Giubilini & Sanyal, 2016, 6; PCB, 2003, 285; Peters, 2003). Seized by a sense of hubris and omnipotence deriving from the mistaken assessment of the depth and accuracy of their knowledge of natural processes, their interventions could lead to irremediable catastrophes (Hamilton in Weckert, 2016, 87; Sandel, 2004). The so-called bioconservatives point out that genetic manipulation could have such deleterious consequences for those involved as to jeopardise their sense of individuality and uniqueness, and hence their authenticity, autonomy and general sense of human existence. Moreover, the idea that genetic manipulation, which should result in an individual improvement and/or in a species-wide empowerment, may actually have positive effects is based on a view of genetic determinism that overlooks other equally fundamental factors contributing to individual maturation. Basically, critics of this particular attitude implicit in human enhancement practices claim that it would be better not to manipulate human nature because the desired results might not be achieved and, in any case, the risks this manipulation might engender are far greater than any possible benefits (cf. Weckert, 2016).

The supporters of enhancement retort with the claim that not everything natural is good for humanity. Indeed, on closer examination, nature seems to have left ample room for improvement, which humanity has always made such strenuous efforts to fill that all its work may be said to consist of identifying the shortcomings and errors of nature and trying to eliminate them. The view that respect and gratitude for nature's work can be translated into a sort of immobility and passive acceptance of its gifts, or at least an abstention from a more or less radical alteration of biological processes or enhancement practices, as Sandel proposes, is not only in contradiction with everything that humanity has always done since its appearance on the face of the Earth, but might even bring about the rapid extinction of the species. John Harris, who has expressed vehement and sometimes violent criticism of bioconservative theories, has no difficulty in pointing out the unsustainability of Sandel's positions through a simple observation: illness and death are also "gifts" of nature and yet there is no outcry when someone seeks healthcare or adopts strategies to postpone death, or simply to maintain a state of health that enables a certain physical and mental wellbeing (Harris, 2007, 112–113). In actual fact, object the bioliberals and the transhumanists, the whole "playing God" issue is based on principles subsequently disproven by facts and on equally mistaken assumptions (Buchanan, 2011b, chs. 2 and 4).

These same assumptions also suggest that human enhancement biotechnologies threaten to undermine the underlying principles of human dignity, a concept that is at least as vague and indeterminate as enhancement (Lanigan, 2009). In a famous essay, Ruth Macklin claims that human dignity serves no real purpose and that its elimination would be inconsequential (Macklin, 2003, 2006) as the evocation of human dignity is nothing more than a slogan and a rhetorical artifice. She then explicitly questions Leonard Kass's view to point out that, apart from a passing reference, dignity is the object neither of analysis nor of problematisation. Macklin may well have a point. The defenders of human dignity do not seem to have very clear ideas of the notion, to say the least, hence it follows that its use in this context is neither constructive nor coherent. Kass, for example, makes an appeal for the uniqueness and individuality of life, its daily concreteness as manifested in the struggle to satisfy and not eliminate the necessities of life. In his view, human dignity lies in the very ability to tackle life's difficulties, pains and illnesses. It is by dealing with adversity that a truly dignified human life is built: an existence that has value because it faces up to the harsh reality of life without looking for loopholes or subterfuges to improve its condition. A good human life cannot be pursued outside its material conditions of exercise and development but only thanks to and in them (Kass, 2002, 17–18).

However, the aim to root the dignity of life in sacrifice, dedication and deprivation comes across as an ascetic and elitist perspective that would be difficult to accept for those who find themselves in a situation of truly having to overcome all kinds of difficulties and would willingly do without them, and for those struggling to survive who find it more pleasant and dignified simply to be able to live. However, things are not as simple and schematic as this suggests.

Hauskeller believes that, regardless of the condition people started out from or the condition they are now in, real strength consists in the relationship they have with their biologically imposed limitations, *whatever these may be*. In popular imagination, which feeds on the pop culture on TV and in comics, the people envisioned by transhumanism are equivalent to Marvel's superheroes: a group of X-men. Such characterisations are capable of things that normal human beings cannot accomplish because they have removed their limitations. In popular belief, a Paralympian athlete suffers the overwhelming weight of their own limitations, which prevent them from doing what a non-disabled person could easily do and is, thus, the opposite of a superhero. Hauskeller maintains that the lesson to learn from the Paralympian is that real strength lies in the refusal to feel limited by the surrounding environment or contextual impediments and in the daily battle with the weaknesses and frailties of the human body. Superheroes have unlimited powers to control nature inside and outside their body, but these unnatural abilities mean that they acquire no merit or no moral virtue. Paralympian athletes, although this view could be extended to all people who have varying degrees of physical disability, reveal their strength not by removing their limitations but by deciding every day that they do not feel diminished by them. Indeed, the more restrictive and compelling the internal and external limits they try to break down, the more deserving their efforts are (Hauskeller, 2016, 7–8).

Transhumanists, and especially Bostrom, in no way deny that human dignity is a value to be defended. On the contrary, they maintain that transhumanism safeguards and valorises human dignity because it strengthens all the characteristics and functions of people who tackle need and necessity and thus give substance and concreteness to the search for this value. However, as will emerge from an analysis of a debate triggered by Bostrom's long essay (2009b) which resumes and develops the analyses of the Hungarian philosopher Aurel Kolnai (1976), this view seems no more convincing than that held by the defenders of human dignity.

Kolnai explains that there are at least two conceptions of dignity. First, it is a quality and a form of excellence: a valid, noble, virtuous person is traditionally considered a "worthy person". It is a quality that is generally acquired to differing degrees, either in relation to one's effort or commitment in a particular field or profession or simply in the behaviour one adopts in everyday life (Bostrom, 2009b 124). In the original Latin sense of the term, dignity is an excellence which, importantly, brings out and perfects non-competitive qualities. Second, there is "Human Dignity (Menschenwürde)", a characteristic that morally elevates human beings above all other living beings (ibid., 125). Given this duplicity of meanings, Bostrom is quite right to point out that misunderstandings can abound where the concept of dignity is concerned (ibid.).

While human rights are related to the latter conception of human dignity, because they are the natural consequence of the specific moral statute of human beings, the social statute of the individual is dependent on the former conception, which is the one Bostrom is interested in and which he claims transhumanism rightly defends (ibid.). Kolnai's characterisation of "dignity as a quality" clearly illustrates why Bostrom claims that transhumanism is better equipped to defend, protect and promote it. Kolnai says that it entails the possession of certain attitudes denoted by calm, confidentiality, composure, measure and discretion in the expression of emotions, and by a certain degree of distance, detachment and invulnerability. Dignity is also manifested through the ability to rise above misfortune and to accept it without suffering its consequences and is, therefore, attributed to people who calmly accept a certain behaviour and who show active empathy towards others (Kolnai, 1976, 253–254). Irrespective of the specific characteristics of dignity, it is clearly not an immaterial and spiritual aspect of the individual, an unalterable moral status associated to rationality or the soul but, rather, a specific human functionality which can be taught and, consequently, manipulated, enhanced, altered and modified. It comes as no surprise, then, that Bostrom thinks transhumanism is sensitive to this conception of human dignity and aims to defend and enhance it. Of course, these abilities need to be rooted in the individuality of the person who manifests or seeks them, but it is clear that they can be enhanced in every way and it is not the means used, whether this be education, natural temperament or medication, that alters their nature, authenticity or coherence (Bostrom, 2009b, 131). The critical feature of this conception is that dignity as a quality is the result of an effort that can certainly be assisted by any means, but which remains fundamentally possible in the context of a struggle with one's inner self and with the outside world. As steel can be tempered in a chemical process to become stronger, the human soul can be tempered by overcoming unforeseen difficulties, and human dignity can be tempered and confirmed through exposure to difficult situations, mourning, pain, illness and so on.

However, critics of the transhumanist movement argue that transhumanism aims to create a new species of human beings, not by modifying the biological substance of the individual but simply by eliminating these obstacles and thus removing the very possibility of producing the qualities acknowledged as characterising dignity so as "to overcome some of our biological limitations". According to Bostrom, although *posthumans* would be different from unenhanced human beings, they would nevertheless act in full compliance with this "humanistic" conception of dignity. In fact, he claims that *posthumans* would pursue the search for dignity in exactly the same way that unenhanced human beings currently do, by struggling every day against adversity and displaying the same qualities that, according to Kolnai, constitute worthy behaviour (Bostrom 2009b, 142). However, it was promptly observed that this attitude would merely be similar to that of a king who deigns to act in a certain way because, as Rubin says, *noblesse oblige* (Rubin, 2009, 147). In Rubin's view, true dignity is not an attitude of simple detachment verging on arrogance, such as that of superior beings who do not have to face any difficulties in order to survive. Even though they present similar characteristics, at least externally, true dignity is forged in the fire of battle, in the face of adversity that cannot be avoided, and not as a means to display superiority, as a king might in knightly jousts (ibid.).

Kolnai's essay presents another aspect of dignity, partially neglected by Bostrom, which seems to support criticism of transhumanism. As already mentioned, Kolnai differentiates between a qualitative conception of dignity and a structural one, which is encapsulated in the form of human dignity displayed in the attribution of a certain type of rights. While the former is the result of personal effort to oppose unfavourable environmental circumstances, the latter is at once present in the individual. In Kolnai's words, the former is the result of a *description*, the second of an *ascription*; the former acknowledges research aiming aesthetically and/or morally to acquire the values and attitudes listed above, the second is the result of an external attribution, which may also be withdrawn, of a certain type of rights and therefore of freedom inherent in the human being as it is here and now.

While this conception of human dignity as an ascription of rights is certainly less precise, less well-formed and may end up being abandoned, revoked or diminished from outside, there is nothing in Bostrom's arguments to guarantee its protection (cf. Buchanan, 2011b, ch. 3). And yet, attention to this conception of dignity would be all the more important because while the former represents a differential characteristic and distinguishes or may distinguish between individuals, this latter conception of dignity as a characteristic universally shared by all human beings is precisely the basis on which to assign equal rights to all human beings.

And it is precisely this conception of human dignity that bioconservatives fear is endangered by biotechnology. As Fukuyama states, the fear is that enhancement will cause the human being to lose its humanity and, above all, relegate it to a sort of purgatory below the threshold for the attribution of a certain type of rights. Enhancement practices, it is feared, could be severely discriminatory, accentuating forms of inequality between human beings and redistributing human rights in a way that would be utterly unfavourable to unenhanced human beings. Defining the essence of human nature entails defining a set of characteristics that guarantee equality between individuals. To say that there is an essence of human nature is to say that all people are equal and have equal rights. In the eyes of Fukuyama, giving human beings dignity—which must therefore be defended—is a principle from which an egalitarian policy should be developed. This dignifying element, which Fukuyama sees as an undefined, undefinable but nevertheless existent X factor, represents a bulwark not against the invasiveness of biotechnology, but against the fragmentation and dispersion of human nature that would make it impossible to attribute identical rights. In addition, there are fears that enhancement is a factor that might also contribute to a widening of the social, political and economic gap in other ways.

Enhancement practices pose a major political problem that is very often underestimated. One of the aims of enhancement is to achieve social benefits, for instance sports doping or the use of drugs initially intended to treat cognitive pathologies and then used in off-label ways to enhance performance in competitive situations (sports competitions, examinations, job interviews, etc.). In such cases, not only would inequalities not be eliminated, but they might even increase due to the rarity, availability and cost of these therapies which would only be accessible to a small elite. Indeed, in order to be truly useful in the social context, it is necessary not only to obtain them, but also to ensure that their availability remains extremely limited. Used in the context of social competition, enhancement therapies only provide an advantage if they create a gap with other competitors, thus producing or aggravating an imbalance. If they were universally available, the equilibrium that their use is intended to upset would be recreated at a higher level. Enhancement thus raises two issues of justice: the first generally concerns the conditions of access and distribution of appropriate biotechnologies, the second their explicitly anti-democratic and anti-equalitarian aims (Buchanan, 2011b, 143–170; Mehlman, 2009).

Nevertheless, it has been observed that considering human enhancement as a zero-sum practice aiming only to increase certain positional or external goods, the result of a personal choice or possibility, means unduly reducing the benefits that this type of practice brings or could bring to society as a whole. In actual fact, says Buchanan, it is the whole of society that has always benefited overall from individual improvements and there is no reason why these new human enhancement techniques should not have the same effect. Buchanan argues that the context in which enhancement practices are acceptable is non-competitive, or at least competitive to an extent such as to be of benefit to society as a whole. The increase in individual differences, and thus the resulting benefit for those who are enhanced, is offset by the existence of what Buchanan calls a "network factor" which also has a positive impact on those who do not directly benefit from such enhancements, in accordance with the second principle of justice proposed by Rawls. Enhancement is not only a positional and, above all, undemocratic good, which, in order to still be considered as such, must necessarily be distributed unequally, but may benefit society overall. However, its alleged benefit can only be assessed if its effects are taken into account in a synchronically and diachronically broader context.

5.2 Therapy vs Enhancement

One of the contexts in which an attempt has been made to define the meaning of enhancement (and, consequently, in which the moral legitimacy of transhumanism and its proposals is at stake) is the discussion on the continuity between medical therapy and enhancement, whatever the latter may be. Attempts to explain what enhancement is have very often been made by contrasting it with therapy and differentiating their respective aims. Proving that a discontinuity between therapy and enhancement is impossible would ipso facto make much of the criticism of enhancement null and void. The objections that are sometimes raised against medical therapy regard the appropriateness of administering certain types of health treatment to certain types of patients who find themselves in extraordinary and extremely complex situations, though the therapeutic act itself and therefore medicine as knowledge and practice have never been questioned or considered morally questionable. Should it prove impossible to distinguish between enhancement and therapy, criticism of the former would also target the latter, and acceptance of the latter would also impact on the former.

Thus, if it were possible to prove that there is a discontinuity between medicine and enhancement, since they are two essentially different practices in terms of the means employed, the desired and achieved ends or the undesired consequences that might result, then the objections that are generally raised against enhancement would be coherent and it would be necessary to discuss and challenge them without being able to count on the support of an alleged continuity with medical therapy. This is the argument of enhancement's opponents, who claim that *medical therapy and enhancement are completely different* and, therefore, while the former are absolutely necessary and legitimate, the latter must prove their legitimacy and moral compatibility with human life.

Enhancement's supporters, on the other hand, argue that, as there is no difference between enhancement and therapy because both aim to improve the individual's situation, the former by enhancing inherent abilities and the latter by eliminating temporary or chronic disabilities, there is no reason why enhancement should be rejected or considered morally illegitimate. In other words, *therapy is enhancement*, and opposing the latter would also mean criticising the former. However, if therapy is enhancement, then the inverse equivalence also holds, that is *enhancement is therapy*, the consequences of which would be extremely challenging. While in the vast majority of cases enhancement tends to improve performance in the actions of everyday life, the effects of this reversal of perspective would profoundly change fundamental features of modern society. Maintaining this equivalence of enhancement and therapy and their presence in all aspects of daily life would lead to the pathologisation and the medicalisation of everyday existence. In other words, the therapeutic nature of enhancement and its trivialisation would make individuals eternally ill and by definition incurable, which would impact social, moral and economic life in ways that can barely be imagined.

For the time being, let us focus on the arguments that have been used to distinguish therapy and enhancement. Although its title seems to suggest a very different aim, the *Beyond Therapy* report presents some totally valid arguments in favour of abandoning the distinction between therapy and enhancement. Not only does it point out that enhancement is an abstract and imprecise term, as already stated above, but also that its aims do not seem to be very far from those of therapy: "all successful therapies are enhancing, even if not all enhancements enhance by being therapeutic" (PCB, 2003, 14), and it certainly cannot be said that the category of "non-therapeutic enhancements" is clear. The WHO definition of health as "a state of complete physical, mental and social wellbeing", suggesting that "almost any intervention aimed at enhancement may be seen as health promoting, and hence 'therapeutic' if it serves to promote the enhanced individual's mental wellbeing by making him happier" (PCB, 2003, 15), seems to deny the very claims made in this report.

One of the first assessments of the difference between therapy and enhancement, made by Eric Juengst, seems to be seriously compromised by the awareness, lucidly expressed by the author himself, of the difficulty in finding an argument in favour of this view. In fact, if therapy and enhancement seem to be in a sort of continuity it is because the values that human beings propose to attain in these two ways are not absolute. It therefore seems logical that a state of health in one context can be interpreted as a form of disability in another, and that therefore a therapeutic or enhancement intervention can be considered alternatively immoral or justified depending on the social, economic or political context in which it is proposed and performed (Juengst, 1998, 31). Clearly, therefore, defining enhancement by opposing it to therapy and stating that everything that is not therapeutic can be catalogued as enhancement becomes an almost impossible task because it considers as absolute that which can only be relative (ibid.). On these premises the possibility for any explanation disregarding the ontology of medicine to achieve a conclusive result must evidently be ruled out.

Nevertheless, Juengst explains that therapy and enhancement can be distinguished in three ways. The first is fairly obvious: only illness, disease and disability are treated (Juengst, 1998, 32). Despite the fact that this potentially includes all sorts of states among those that need therapeutic care, a valuable therapeutic indication is provided: as long as the human body functions without any major imbalance or disruption, there is no point in trying to repair it and it is better left alone. However, apart from the fact that this view accepts the tendency of contemporary medicine to seek out diseases even in the absence of symptoms and to prevent the onset of a pathology, this "definition" of disease clearly implies a social and evaluative aspect that prevents a line from being drawn between enhancement and therapy. Juengst adheres to a notion of health that implicitly corresponds to a state of normality, a highly problematic notion that is stigmatised in *Beyond Therapy* (PCB, 2003). It certainly does not help to say that *normal* functioning corresponds to the performance of the functions for which the organ in question is *naturally* responsible. The intuitive perception of thresholds of normality clashes with problems that cannot be resolved, because the definition of these thresholds owes as much to the ecological context as to the social one and, furthermore, the difference between natural and non-natural is just as difficult to define as that between therapy and enhancement.

Thus, Juengst has every reason to say that the rule of not fixing what is not broken, which he feels is valid, is rather difficult to apply. Especially so, as identifying and subsequently enforcing a boundary between enhancement and therapy is complicated by preventive medicine, which does not treat current "dysfunctions" but steps up the body's defences against possible future attacks and acts on what is not "broken": as such it can immediately be regarded as an enhancement practice (Juengst, 1998, 33).

The second criterion points to "luxury" treatments within enhancement, while therapy simply aims at restoring normal, standard functions of an organ or organism. However, the functional normality of an organ is an extremely delicate concept to handle and thus also this distinction immediately loses its applicability.

The third criterion is actually a form of abdication from any ontology: Juengst believes that the difference between therapy and enhancement has to do with a social construction reflecting the social and medical relativism it engenders. Therefore, it is the professional values that medicine aims to follow at one particular time that supply criteria to separate therapy from enhancement (ibid.). The only result obtained in this case is the liberation of medicine from any constraint, leaving doctors free to intervene in the way they consider most appropriate. However, the main objection that could be raised against the latter view is that, by reducing it to a private issue, the distinction between therapy and enhancement becomes utterly unusable.

Other arguments have been put forward to tackle this inability to distinguish between enhancement and therapy. For example, it has been suggested that defining the type of suffering that requires medical intervention may be of help in this respect. However, the same difficulties are encountered, as even cosmetic surgery, which a priori is not a strictly therapeutic operation, may be regarded as such in the case of an individual who has been disfigured. In this case the suffering that the doctor is treating is not so much physical as psychological and social: by restoring an "ordinary" appearance, the individual in question would be able to resume a normal life from all points of view. It might be objected at this point that it is not up to the doctor or surgeon to eliminate the causes of psychological and/or social "disorders" and that they should only treat physical pathologies so as not to overstep the bounds of their professional competence. This argument should be accompanied by considerations of distributive justice and a conceptual arsenal aimed at ensuring equal opportunities for all individuals whatever their situation, which clearly cannot be discussed here.

Another possible argument against the impossibility of separating therapy and enhancement is that medicine has legal and not only technical limits, which it must not cross. For example, as stated in the French *Code of Ethics*, while every doctor is obliged to assist any person in need of treatment, they must restrict "prescriptions and actions to what is necessary for the quality, quantity, safety and efficacy of care" (Goffette, 2009, 57). However, here again rhetorical acrobatics that allow care and enhancement to be confused or amalgamated claim to find justification in the very dictates of the rule: what does "efficacy of care" mean? At what point does therapy become efficacious or stop being efficacious because it has achieved its purpose and become something else? What does the quality of patient care entail? The answer to these questions cannot ignore the use of such notions as health, illness, pathology and normality which, of course, result in numerous contradictions.

A further argument that can differentiate between medicine and enhancement relies on the diversity of the doctor–patient relationship in the two contexts. A situation in which a doctor has a patient *needing* treatment can be defined as therapeutic, while a request for enhancement can be interpreted as a patient *seeking* treatment. Although this difference between *needing* and *seeking* patients seems legitimate, at least at first sight, it does not appear to stand up to more thorough scrutiny. One and the same pathology may impact patients differently: some pathologies are more disabling than others and put patients in a state of need, while others simply make them less able to perform a certain number of everyday actions. These differences can also be extended to situations in which different individuals are affected by identical pathologies with the result that their treatment could legitimately be regarded as both enhancement and therapy. Establishing the boundary between two individuals hoping to receive the same type of treatment for different reasons seems to be an extremely delicate task and is unlikely to entail *solely* the evaluation of objective elements.

The conclusion of this brief overview is that therapy and enhancement are similar practices as no difference can be established between them and, consequently, enhancement is legitimised. This deduction is made through the rejection of the contrary argument, that is because it is impossible to prove that there are differences between therapy and enhancement. As a result, it must be admitted that transhumanist enhancement practices cannot be considered illegitimate because they merely reformulate and repropose what humanity has always tried to do but with other means: to lead a good life while staying healthy, that is to achieve "complete physical, mental and social well-being", if not fully then at least in part.

However, if therapies are enhancing, regardless of how this is interpreted, will the opposite not also be true, that is that enhancements are therapeutic?

Many objections have been raised against this eventuality, which opens up disturbing and dystopic scenarios. The first of these is rather problematic in that some enhancements are non-therapeutic as they are not achieved pharmacologically or medically. For example, coffee, tobacco, sport training or other forms of body care are all products or techniques that are clearly intended to improve performance and general wellbeing, but they are certainly not therapies in the proper sense of the term. However, if the use of a drug to improve concentration is legitimised by the claim that its effect is similar to that of caffeine, which people consume without questioning its morality, does it not follow that the opposite will also be true, that is that caffeine use is therapeutic? If this is the case, and there is no obvious reason why it should not be, does it not follow that all enhancement practices are therapeutic, regardless of the means used?

This is a somewhat delicate question. If the answer is yes, then daily life might become entirely pathologised. The reduction of enhancement to therapy would end up with a conception that considers people as being fundamentally ill and always in need of care and therapy. Moreover, there is clearly nothing scientific in this outcome, which is the result of a social and intersubjective construction entirely generated by the structure of modern society and its civilising process. In addition, the investigation into transhumanism's moral legitimacy should not fail to carefully evaluate the possibility that transhumanism is an outcome of medicine. This is not such a remote possibility and it could be considered from two different perspectives. Transhumanism might derive from medicine for two reasons: it might be its *natural* extension, or it might be the effect of a *problematic* extension of its aims. The attempt to sanction transhumanism by highlighting the goals it shares with medicine would end up showing either that transhumanism is the poisoned fruit of medicine, its illegitimate child, or that by changing and modifying its means and ultimately also its aims, medicine could not but generate transhumanism. In this way, transhumanism may not be the fruit of wicked minds but a legitimate consequence of a distortion of the aims of medical therapies or the inevitable outgrowth of the ongoing perversion of medicine. In the former case, that is that there is a misunderstanding of the aims of medicine that justifies and legitimises the aims of transhumanism, then the baby (medical therapies) will not have been thrown out with the bath water (transhumanism). Whereas, if the advances of medicine could not but lead to transhumanism, then this movement is, ultimately, fully consistent with the time in which it arose and rightly manifests its implicit and explicit tendencies.

5.3 Medicine and Transhumanism

In order to at least begin to sketch out an answer to this question we need to understand whether the history of medicine is a continuum from a theoretical point of view or whether it has experienced discontinuities and, if so, what they are and what they entailed. Even in the absence of an in-depth knowledge of medicine, there can be no doubt that it has evolved over time as a result of technological improvements as well as a deeper understanding of natural processes. This superficial response is clearly insufficient, however, as it becomes necessary to understand whether the changes medicine has undergone have been such as to make a movement like transhumanism the inevitable outcome.

Ancient medicine, say up until early modernity, aimed to achieve *restitutio ad integrum*. Based on the analogy between the macrocosm of nature and the microcosm of the human body, human health manifested itself in the balance of bodily humours, just as the health of nature was expressed by the proper balance of its different elements. The important characteristic of this conception of human health was that the balance of humours could not be enhanced and had only two states: all or nothing, imbalance or balance. Therapy made it possible to go from *dyscrasia* to *eucrasia*, but the latter could not be enhanced and the best that therapy could achieve was *integrum*, "integrity" (Wiesing, 2008, 9–10). In other words, the best was integrity and integrity was the balance of the different organic fluids, or to use a political metaphor, health was a situation of isonomy, while illness was produced by the sovereignty of one humour over the others. This conception of health was based on a view of physics that saw nature as divine work and left human beings only the possibility of imitating it but certainly not of intervening in its mechanisms in order to improve it. Ignorant of the processes of nature, the mysterious work of a divine intellect and as such inscrutable to the finite and limited human mind (which was also a divine creation), humanity could only imitate the divine work and restore it when internal or external causes put it in danger, but certainly not transform it. In early modernity, the exponential development of science radically changed this conception to leave more room for human creativity based on an enhanced understanding of nature. In actual fact, an even greater change impacted the anthropological paradigm: from the seventeenth century onwards the human being was no longer considered as a microcosm reflecting the natural macrocosm, but first as a machine and later as the result of the evolutionary process. These two models led to the abandonment of a concept of health based on balance while also leaving ample room for the idea of their possible enhancement (Wiesing, 2008, 13). The scientific revolution of the seventeenth century also brought about a change in the means and aims of medicine. Nature to be imitated is replaced by nature that can be changed and improved on: no longer the definitive work of a divine intellect and as such deserving of respect, but a set of random processes that can therefore be improved (ibid., 13–14). The immediate consequence of this view is that the aim of medicine is no longer to restore a lost balance, but the transformation of natural processes in order to make them better, more effective, less costly and above all more durable in order to achieve a *transformatio ad optimum*.

Summarising what has been said so far, the bewilderment engendered by transhumanist enhancement and the inability to explain why some enhancements seem legitimate and natural while others seem intuitively wrong and immoral, are the result of a change in the aims of medicine (from *restitutio ad integrum* to *transformatio ad optimum*), in turn brought about by the scientific revolution and perhaps by the very project of modernity, the consequences of which have not all been fully evaluated. Beyond all the factual issues, there is a conceptual difficulty inherent in the development and definition of the aims of modern medicine that makes it impossible to establish this boundary. The aims of medicine changed with the rejection of Aristotle's philosophy of nature and the medical practice of Hippocrates and Galen, which blurred the concept of health and the definition of sickness. What was lost was, very concretely, both the ability to distinguish health and sickness and a criterion establishing a difference in value of these two states. In other words, contemporary medicine finds it difficult, not to say impossible, to explain how sickness is "something other" than health, or pathology is in discontinuity with physiology, as can be seen from the fact that the two paths followed in order to achieve this (one quantitative, the other qualitative) have not offered convincing results. However, the conclusions of these attempts are of interest because they highlight a problematic but not frequently discussed aspect of transhumanism.

First of all, stating that illness is a state of health in which something is lacking—for instance, the oxygenation of the blood in the lungs—or something is in excess—such as an inflammation of the gallbladder, causing it to secrete too much bile—and thus referring to a quantitative criterion, has not been very helpful. This type of definition is open to too many exceptions, as represented by a state of health compatible with quantitative variations in human physiology that make it impossible to identify and isolate the two states in this way.

Attempts to indicate a qualitative discontinuity, as proposed by Canguilhem, obtained no better results. Having observed the inconsistencies into which his predecessors had fallen, Canguilhem aimed to find a criterion to measure the qualitative difference between health and sickness in the normative force of life. His reasoning set out from the observation that a sick organism, or an organ that has suffered trauma, reacts to the alteration it has undergone by trying to perform the same actions as before in the best possible way. If the condition results in an overall disorganisation, the organism tries to minimise its effects through an optimal organisation that takes into account the new situation created by the sickness. Canguilhem gives the example of a person suffering from arthritis of the knee who changes the position of their leg in order to carry on performing the same movements without the pain. The new position is *defective* only in relation to all the possibilities that the knee had prior to the onset of arthritis, but from an anatomical point of view there is no change because the knee can do the same things it did before, even if it has become normatively less *flexible* (Canguilhem, 1991, 183–184). Health and sickness are not in a relationship of more and less but become two different solutions to the same problem posed by the environment: the only substantive difference lies in the type of solution put into practice, which can be more or less elegant and more or less effective. In this way, not only are all the differences between health and sickness annulled, but sickness is no longer a different state from health because the organism can *never* find itself in a dysfunctional state. Nevertheless, Canguilhem thinks that it is possible to separate health and sickness. He claims that a healthy organism differs from a sick one not only in the elegance or the efficacy of the solution to the life problems it faces, but also in the number of solutions it can explore and the different life force with which it projects itself into its environment. An individual is sick because they cannot admit as many norms as they could before their sickness: in other words, their sickness is denoted by a loss of regulatory capacity or rather of adaptability to the surrounding environment. In maintaining that sickness is not a quantitative variation of the state of health, Canguilhem acknowledges that it is a new dimension of life which is nevertheless characterised by a reduction in the individual's capacities: in other words, the abnormal is such only in relation to context. A healthy man is potentially identical to himself when he is ill; it is the expressed norms of life that are different; and thus the pathological is marked by a normative poverty of the sick organism compared to a prior self (Canguilhem, 1991, 144). This leads on to the discontinuity between pathology and physiology, which may also have a much broader application as a pathological state may allow an individual to adapt to a specific environment with a series of extremely restricted adjustments that, above all, reduce their potential to adapt to other environments, as Canguilhem observes. In short, the life expectancy of a sick individual has shrunk due to the effort they need to make in order to actualise and define new life norms. The healthy individual is *adaptable* while the sick individual has *adapted*: *adaptation* and *adaptability* are indirectly proportional, with the consequence that an individual who is perfectly adapted to a given environment has partially lost the ability to adapt to other environments.

What consequences does this have for an interpretation of transhumanism? Transhumanism sees itself as a philosophy of power and great health: the enhancements it proposes and fosters are intended to eliminate all the shortcomings of the human body. The assumption is that human beings are affected by pathologies, that is by restrictions of their life force which lock them in a limited number of ecological niches. Healing is seen as an indefinite extension of the environments to which it can adapt: potentially, a healthy body is indefinitely adaptable. However, this certainly attractive view has at least two less acceptable features. The first is that adaptation techniques, the therapies according to which transhumanism intends to heal human beings, would occur at a rate that would soon become unbearable and would transform the human body into a constantly expanding machine. The second, as already mentioned, would be a pathologisation of daily life that would reduce society to a field hospital and existence to a continuous search for enhancements whose meaning would no longer be clear.

After this long examination of the attempts to unequivocally and unambiguously separate therapy and enhancement, an instant conclusion that can be drawn is that they encounter insuperable difficulties and are probably doomed to failure. The reason for this failure, albeit expressed differently, lies essentially in the impossibility of defining an ontology of medicine (which would actually be an ontology of biology): any attempt to separate therapy and enhancement meets with insurmountable obstacles if there is no will to address metaphysical and ontological issues that are currently outside the scope of doctors' and especially biologists' interests. But is it still possible to evade the aporia to which the ontological void gripping biology delivers us? Perhaps, but on condition that we abandon any naturalistic conception of medicine and consider sickness as a structurally social phenomenon constructed on and related to the values of society. Inasmuch as medicine contributes to the establishment of social and moral values, it too must become aware that it is an entirely social practice with all the consequences that this implies. The first of these is that it will be necessary to find a lowest common denominator for social and moral values to which medical practice must somehow be legitimately subordinated. In this sense, only acts that contribute to the promotion of that minimum value can be considered therapeutically relevant.

5.4 The Instrumentalisation of the Body

The transhumanist movement is rooted in two fundamental assumptions that are both sides of the same coin. The first is a negative anthropological conception. The desire to enhance the human body is in fact the result of an implacable diagnosis: although the human body is a wonderful creation and a machine that works reasonably well, it has many defects and therefore ample room for improvement. The second is an awareness of the intrinsic dynamism of life that, the transhumanists claim, justifies enhancement aimed precisely at changing the biological structure of the human being, by following in the footsteps of nature or, rather, by enhancing evolution as in the title of the book by John Harris. Some of the consequences of this acknowledgement are rather paradoxical and aim to dodge the criticism directed at transhumanism by removing its object. In particular, unexpected perspectives are opened up by the idea that our shortcomings and our efforts to eliminate the natural imperfections of our biology make us "natural-born cyborgs", that is beings whose nature is ontologically hybrid. This means that if mental states or any other human production are the result of cooperation between organs and tools, it would be legitimate to affirm the existence of a two-way equivalence whereby a machine is an organ and, vice versa, an organ is a machine (Canguilhem, 1989). This collaboration, which puts organs and tools, the natural and the artificial, the biological and the technological, on an equal footing, produces profoundly alienating effects on common sense, as will be shown below. In short, the anthropology of transhumanism generates two images of humanity: one negative, representing it as disabled, and one positive, portraying it as a cyborg. The foundations of transhumanist thought are thus not represented by the awareness of human weakness or by the possibility of enhancing the human being, but by the *simultaneous presence* of these two elements that cannot be isolated without bringing down the entire edifice of their justifications: if we can become stronger, more intelligent, more able and more long-lived, it is because we are currently lacking in these qualities, it is because we are inadequate and aware of the weakness, fragility and inadequacy of the human body. There is no reason to look outside the mechanism that holds these two elements together, because ultimately there is nothing outside this interplay of the best and the worst that reciprocally balance each other.

Understanding the reasons why these two representations can coexist must necessarily start from the implicit assumption of these two aspects of the wild anthropology of transhumanism, according to which the body is a tool and, indeed, "that very first human tool" (Gray, 2000, 3). But is this really true? Is the body ontologically just a tool? And, moreover, is it morally legitimate to reduce the body to a tool that the subject, the intellect, the individual, the soul, the mind or the ego use? And even if this were the case, what impact would its modification have on the whole of which it is part? Can it simply be reduced to a formal change?

The transhumanists' answers to these questions are generally positive. Transhumanists (More, 2013a) and other bioliberals (Buchanan, 2011b, 156) adopt an expression used by Richard Dawkins (1986) in stating that the human body is the product of a blind watchmaker who, because of his blindness, could only do imperfect work that deserves to be improved. It is therefore right to show respect and gratitude to nature for its work, as Max More does in his *Letter to Mother Nature*, but at the same time it does not seem wrong to point out the weaknesses and shortcomings that the human body presents and propose to eliminate them (More, 2013c). What would be wrong in completing a work that, although admirable, is defective and unfinished, lacking and imperfect, like the human body, for example by increasing and enhancing cognitive abilities, eliminating disease and death, or even broadening the scope of its perceptions? A priori nothing; indeed, it would be a meritorious work.

This recommendation is based on the analogy between the human body and a machine. After all it is said that the human body is like a machine that could work better and it is therefore worth thinking about how to repair, improve or perfect it. The very possibility of thinking that the body, while not imperfect, is certainly in need of improvement, is based on an analogy with a machine. Although it is not entirely mistaken, this analogy is certainly incomplete. Machines or tools are artefacts built for a specific purpose, whatever that may be. Obviously, whoever invented and built the clock had a precise purpose in mind—to measure time—so a defective clock is the one that does not perform its functions properly. Hence, a machine that needs to be repaired, improved or perfected is a machine that does not fully accomplish the purpose for which it was built. Leaving aside the analogy with the machine for now, an evaluation of the body, whether positive or negative, cannot be carried out without considering the objective it has to achieve: that is we can say that the body is defective if we know what purpose it serves. But we do not know this and none of the authors in question is mindful to tell us. It is evident, therefore, that a body that first of all has not been created by anyone and that has no end to achieve cannot be said to be defective: not having an end to achieve, it will be neither perfect nor imperfect. This does not detract from the fact that it is good to want to eliminate disease, pain, suffering and death; the problem is that the reason for this meritorious intent, which is based on an alleged incompleteness or incapacity of nature, seems utterly mistaken and in any case misleading.

Also because (and this is the second reason to eliminate any reference to evolution and avoid stating that enhancement is basically nothing more than an improvement of the evolutionary work of nature, i.e. rejecting the idea that the purpose of transhumanism is to enhance evolution) if we cannot explain what function the human body must perform (and we cannot), then everything can potentially be improved and perfected because everything is defective and lacking, which has much further-reaching consequences than a rejection or acceptance of the aims of transhumanists.

Finally, in the face of an assertion that equates the body to a tool, indeed the fundamental and primary tool, it would appear trite to ask whose tool it is or who the intended user is. Obviously, the answer to this question would risk introducing a form of dualism that the scientism of transhumanists would be utterly unable to accept. However, a materialistic and more precisely methodologically mechanistic perspective makes it possible to avoid dualism and at the same time to affirm the instrumentality of the human body. But is it possible to justify the evident reductionism of transhumanism that sees the human body as a machine? What difficulties are inherent to this theory? Can they be overcome? Clearly, this is not the place to review all the figures and attempts to portray the human being as a machine. However, it would be possible to use a statement by the author of the article "Mécanicien" in Diderot and D'Alembert's *Encyclopaedia*, as a signpost on the path to the clarification and perhaps the rejection of the analogy between body and machine: "the animal body, and consequently the human body, is held to be a real machine; that is, a compound body whose parts are made of such a kind of matter, figure and structure that, through their connections, they are likely to produce certain effects for a predetermined purpose" (Israel, 2004, 67). Is this correct? Is the specificity of the human body truly understood and represented in this analogy with a machine? It is doubtful.

Let us start from the observation that there are at least two principles on which mechanistic reductionism of the human body is based. First, reductionism is based on individualism, in the sense that the body is nothing more than an aggregate of simple, indivisible elements such that its behaviour is the result of the action of its individual elements. Of course, this atomistic individualism is based on the principle that the whole is simply the sum of the parts, the validity of which has been positively contested. Indeed, it has been observed that many dynamic systems display a behaviour that cannot be explained as the mere sum of the behaviour of their individual elements. In an attempt to preserve the principle of atomistic aggregationism, it was thought that the system features not directly attributable to the behaviour of the single elements, can however be indirectly referred to the elements that make up the machine because they emerge from their interrelation or, in other words, are emerging qualities of the system. Here, too, criticism has not been lacking since the process that would have led to the emergence of one quality and not another remains incomprehensible. In fact, it seems more sensible to abandon the idea on which the principle of atomistic individualism is based, that is that the body is a closed system. The statement that the whole is not the sum of its parts, that is the body is not only the compendium of its organs whereas the machine is the compendium of its assemblies, aims to argue that the overall behaviour of a system cannot be described and explained only in terms of the behaviour of its individual parts, but it is necessary to refer to something external, that is, to something that is already given as external, i.e. the context in which the system operates (Israel, 2004, 101).

Just how rooted these two aspects of reductionist individualism are in contemporary biology and how important they are for transhumanism is evident from enhancement techniques that focus on the improvement of the individual parts, organs, cells or neurons of a person. For example, the possibility of rejuvenating individual cells is presented as proof of the possibility of defeating old age and even death. Clearly, if the correctness of reductionist individualism were to be denied, then it would not simply be this or that enhancement that would lose its effectiveness, but the very possibility of thinking that the enhancement of an individual organ or function could actually improve individual life would fall. Even more radically, the theoretical foundation of transhumanism, its very *raison d'être*, would be lost.

Secondly, when the subject is reified on the basis of its macro-behaviour, it is regarded as an agent that is not only rational but also able to act according to functions of utility, that is as an agent capable of always identifying the best way to maximise its own interest and its own wellbeing. Applying this principle to transhumanism is fundamental because it makes it possible to assume that, faced with different enhancement possibilities, the subject always knows where utility lies (wellbeing, happiness, interest, etc.) and can choose the best way to achieve it. However, as people are very frequently (indeed almost always) unable to irrefutably express and justify a preference, the hypothesis that "it is equivalent to admitting that the individual has a perfect knowledge of all the possible options and a doubt-free capacity of choice", or rather the so-called "hypothesis of the unlimited foresight of the individual in question" (Israel, 2004, 104), is contradicted by the facts. This hypothesis is also mistaken from a logical viewpoint: how could an individual, who is imperfect by definition, know what constitutes an improvement in their situation? They lack two elements: a vision of the future that would allow them to modulate their improvements in an enhancing way, and knowledge of the necessary perfection to be achieved. Clearly, only through an analysis of the body with the external world might these two data items be derived and, regardless of the possible acquisition of these data, the conclusion is that their behaviour cannot be explained exclusively in terms of a closed system in which the whole is the sum of the parts.

Finally, let us remember that these two principles of mechanical reductionism of the body, that is methodological individualism and the principle of human agency aimed at maximising one's own interest, are not only scientific statements but constitute the theoretical basis of liberalism. Hence, this simple analysis of the consequences of the body-machine analogy makes it possible to deduce the coherence and compatibility of transhumanism with political liberalism. It will be the task of the following sections to point out all its features and consequences.

5.5 An Anthropology of Disability

This specific vision of the human being, embedded in the debate on enhancement and the enhanced individual, raises the issue of disability and, more specifically, the idea that ultimately all people are disabled and unsuited to the environment in which they live. Although disability is manifested in many ways, for transhumanists it is an innate constant of man. By implicitly extending Max More's observations, Savulescu, Sandberg and Kahane claim that everyone is disabled (2011, 14). Harris, on the other hand, also sees disability as a "harmed condition" (2007, 180) that would never be the object of a rational choice, since it would unduly limit the possibilities of having a good life or achieving one's goals (Harnacke, 2013). Although we can agree with Harris that no one would choose to be disabled, we must also recognise that a life can be good and worth living even if it only has a relative degree of functionality, as Hauskeller suggests and as a careful observation of people with disabilities points out. A different and perhaps more appropriate conception of disability would help to put the aims of enhancement back into perspective, because "less does not necessary mean worse, and [...] perhaps more may not equate to better" (Eilers et al., 2014, x). The inherence of disability in human nature may give rise to two completely different readings and produce substantially opposite outcomes.

First, disabilities are the subject of a medicalising paradigm (Coeckelbergh, 2013a; Charis Hall, 2017; Flanigan & Price, 2018; Eilers et al., 2014). More precisely, people are said to be disabled because they have less functional abilities than those typically possessed by humans in the same situations and for the same functions, and this diminished functionality is the result of a biological anomaly (Satz & Silvers, 2000, 173). Regardless of their nature and type, these disabilities must be eliminated because they lessen "the value of our lives" inasmuch as they make it more difficult for us to lead and achieve a good life, or at least what the disabled person thinks a good life is for them (Savulescu et al., 2011, 14). They are pathologies in need of a therapy of any kind that will restrict and reduce the negative effects on the disabled person in question. Clearly, the medical paradigm that considers disabilities as defects, limitations and, ultimately, pathologies to be healed and corrected is justifiable on the basis of the conception of the human body as a closed system, the difficulties of which have already been highlighted. With this awareness, that is having highlighted the problems inherent in the reductionist vision of human existence on which this theory is based (Winance et al., 2015), we are able not only to relativise the positive impact of biotechnology on individual wellbeing, but also to warn against considering individual enhancement as a panacea.

It is also with regard to the above comment that the medicalising paradigm of disability is not universally accepted and indeed another conception of disability is very often preferred. According to another more complex model, disability is the result of the interaction of different factors and not the failure of a biological structure isolated from its context. As Tom Shakespeare and others have observed, one should not indulge in the temptation to make disability a simple medical problem by decontextualising the various organic functions of the body (Shakespeare, 2014). It may be true that a person with a motor disability has a physical deficit, but it is equally clear that it is their personal context that makes it difficult, if not impossible, to move, thus exacerbating a functionality that in another context might have better achieved its purpose. According to Shakespeare, therefore, it is wrong to think of disability as a medical and individual problem, since it is the result of the interaction between an individual and the context in which they find themselves. To think of eliminating a disability through intervention in the individual is the result of a reductive and fundamentally mistaken conception of the disability itself, because disability might at least be reduced, if not eliminated, in a considerable number of cases through interventions on the environment. In this case, the body is considered an open system that interacts with the environment: its performance is the result not only of isolable modules within the body but also of the interaction of these modules with the surrounding environment. Consequently, when embracing the social model of disability, it is possible to eliminate or reduce its impact on disabled people, with forms of social enhancement that act on the context in which the subject is located. As Cabrera writes, social enhancement is "any intervention that augments or improves an individual's capabilities set with the aim to enable and empower them as active members of society, without directly changing the biological reality of individuals" (Cabrera, 2015, 93).

This social model of disability has at least two advantages: it avoids the stigmatisation of individuals who are not able to provide the same performance as that statistically observed in individuals of the same species. Moreover, it partially or fully shifts the pressure to improve onto extrinsic and contextual elements, thus avoiding a pathologisation of people.

Of course, transhumanism tends to favour the medical model of disability, which offers an unreserved justification of human enhancement. Adopting the medical model of disability entails a view that more or less everyone is prevented from achieving a good life and that the only valid intervention is one that acts on the body and not on the environment. In other words, medicalisation has the more or less undesirable side effect of a progressive invalidation of the politics.

However, the pathologisation of disability encounters a logical difficulty: how can a good life plan be drawn up outside the context in which one finds oneself? This objection can be formulated more precisely using the concept of adaptive preference: the production of desires, whether for a good life or a certain kind of performance, seem to be the result of a compromise between will and reality. In other words, there are no absolute desires, values or preferences, but only those relative to a given context, with respect to which they must be measured. If this is so, then disabled people (and thus, according to transhumanists, all human beings regardless of their specific biological condition) also express the desire to lead a good life that can only be relative to the condition in which they find themselves and that is as good and valid as any other life plan. Otherwise there would be absolute values and conditions—the origin of which would have to be identified—that everyone would try to achieve and that would be universally desirable because they are rooted in nothing other than an abstractly evaluated and no longer concretely situated human nature. Obviously, enhancements could result in a hyper-competitive situation: the objectives might in fact be exclusive and therefore open only to those able to achieve a certain type of performance. In this case, enhancements would be positional goods that, by definition, should be restricted only to a part of the population in order to be able to bring benefits.

In short, it may be true that humanity has always tried to improve itself, but transhumanism considers that the most valid strategy to pursue (and perhaps the only feasible one) is the correction or elimination of human weaknesses. Traditionally, humanity has tried to improve its condition in two ways: the first is *eliminatory*, while the second is *corrective*. The first aims to eliminate the obstacles preventing the full deployment of human potential and is pursued by intervening fundamentally in the environment, in the ecological niche in which people act; the second aims to correct the functions and the organs that produce them and consists essentially of therapies designed to re-establish the status quo. Up until the second half of the twentieth century, enhancement interventions were balanced in that they acted both on specific ecological niches—for example, all forms of social hygiene aimed at improving the environment and adapting it to the specific characteristics of the individuals who lived in it—and on the biological matter of individuals to enhance their organic characteristics and potential—for example, vaccines.

Transhumanism has no doubt that these two strategies are equally valid but, in line with the above observations on the adoption of a medical model of disability, argues on the one hand that interventions in the ecological niche are not fully satisfactory, and on the other hand that the tools used are inadequate because they are partially or totally ineffectual and in any case too slow and cumbersome (Agar, 2014; cf. Doyle, 2018). In other words, in a substantial continuity with traditional human enhancement practices (at least with regard to the ends), transhumanists propose to use a completely different strategy that they claim would have an incomparably greater efficacy. Instead of intervening in the environment, they seek to use genetic manipulation in order to eliminate the human body's internal obstacles that prevent the full deployment of an individual's potential in a certain environment. In view of how mobility could be improved, for example, it seems clear to transhumanists that genetic manipulation aimed at muscular and cognitive enhancement or the application of a prosthesis with sonar capability would be much more effective, faster, cheaper and more satisfactory than a policy aimed at eliminating architectural barriers. However, since this proposal is based on an atomistic and closed model which equates the human body to a machine, the inconsistencies of which have already been highlighted, it seems highly unlikely that transhumanism will actually achieve its stated aims.

However, not only is it possible that transhumanism promises things it cannot provide, but it may also have negative effects for the survival not of humanity as it is but of the enhanced individuals or the new species of human beings that would supplant them.

The awareness of the intrinsic dynamism of the human species has become the theoretical basis to justify all interventions in our inadequate biological structure. It is on these foundations that enhancement or perfection have become the watchwords, not only of humanity's most recent history but of the project of modernity itself ever since the writings of Francis Bacon. Modernity has not only freed time from the shackles of static contemplation, but it has set it on the path towards the future: if not a goal, then certainly a meaning. The awareness that it is not in the nature of the human species to be what it is has been supplemented by the idea that this continuous movement can be guided or governed (a term that in modernity is rich in significance and widely used more or less implicitly) in a direction that can be controlled by giving it a meaning, and thus stealing it from the causality and uselessness of abderitism which, on seeing a rise after a fall, "produces no more effect than if the subject of them had remained stationary in one place" (Kant, 1991, 178).

Buchanan's complaint, namely that the level of the debate on human enhancement is somewhat frustrating because knowledge of the laws and the mechanism of evolution is very superficial (Buchanan, 2011b, 6), aims to remind us that the theory of evolution provides a framework that forcefully legitimises interventions in human nature. The upshot of this call to remain within the framework of the theory of evolution is that if the nature of humanity is unstable and mutable, then all criticism of enhancement tending to defend an essence of human nature from disrespectful manipulations is based only on the imagination of philosophers poorly educated in the biological history of humanity. The repeated appeals not to alter human nature demonstrates at least a poor understanding of evolutionary dynamics. However, in spite of what Buchanan says and in consideration of the sheer impossibility of actually taking into account the theory of evolution when speaking of enhancement, it is also possible that this theory, invoked to eliminate unfounded criticism of human enhancement, may not only be found insufficient to justify enhancement practices, but might also represent a highly problematic element for enhancement supporters themselves. As Norman Daniels has argued, it may well be true that the belief that enhancement might alter human nature is much less well-founded than it might seem at first sight, but it is equally true that it is not entirely fanciful (cf. Bostrom & Sandberg, 2009; Daniels, 2009).

Setting aside these difficulties, the transhumanists forcefully underline that if *Homo sapiens* has become taller, stronger, more intelligent and more long-lived under the pressure of its constantly changing habitat, it seems quite legitimate to continue and consolidate processes that are imprinted on the living being's biological nature. Like every other species, *Homo sapiens* is in constant flux, which also means that it does not have a definite and definitive form that somehow encapsulates its essence. Subjected to the need to always be different from itself in order to survive, it seems almost logical that humanity should take charge of its own alterations and direct its course by eliminating random chance, which may yield positive results of course but can also have more than negative consequences. These enhancement interventions would thus not be very different from giving evolution a helping hand to accelerate it and steer it in the right direction (Sorgner et al., 2013). And if it is necessary to help and govern evolution, it is because transhumanists argue that human beings, like all living beings, lack or need something and can be enhanced.

Of course, this is not a new theory as it can be found practically throughout the history of philosophy and has appeared with some frequency since the eighteenth century. However, unlike the transhumanists, philosophers from Herder onwards did not limit their writings to highlighting human deficiencies and frailties but explained that, in reality, these weaknesses underpin the strength of humanity. In fact, the lack of biological characteristics that are common throughout the animal kingdom is matched by the presence of capabilities that have found a way to make up for these deficiencies by inventing and creating the means to increase or create functionalities that were biologically lacking in humanity. For example, the genetic heritage of some animals has provided them with organs that make it possible to stabilise the internal body temperature, thus providing effective protection from cold or heat and enabling them to survive in environments with extreme temperatures. At the same time, however, the stabilisation of these organs makes them unable to adapt to environments other than those in which they live: a polar bear would not find it easy to adapt to an equatorial climate and, vice versa, an elephant would not be able to survive in Alaska. Human beings suffer in extreme environmental conditions because they lack adequate organs and have, therefore, had to devise solutions to protect themselves from cold or heat. But this same lack makes humanity adapt and be adaptable everywhere: its deficiencies are actually its strength because it is suited or can adapt to all weather conditions, so much so that the species has spread all over the globe. With regard to this initial level of superficial considerations, it must be said that *Homo sapiens* has successfully adapted to the environment by modifying it to some extent but above all without altering human nature—a modification that might have jeopardised humanity's adaptive flexibility. In order to defend itself from the cold, humanity tends to live in warm environments, heating them in a variety of ways, such as the use of coverings or by means of heat production. However, on abandoning an ecological niche that requires these heating techniques, the human body can easily adapt to an environment that has an equatorial temperature. Understandably, this adaptability comes from the flexibility of the human body. It is very likely, however, that intervening in human biology would risk blocking it in a conformation that would perhaps be optimal for one ecological niche but would not allow it to adapt to different environments. In conclusion, the attempt to enhance the human body considered as an imperfect, incomplete or needy work of nature would therefore not only risk not improving it at all but perhaps making it even weaker and more vulnerable.

5.6 Becoming a Cyborg

One of the central themes of transhumanism is the conception of the relationship between humanity and the environment. Traditionally, adaptive pressure is entirely exerted on the environment, which must be shaped to the needs of its human inhabitants, whereas transhumanism overturns this perspective, claiming that humanity must be modified in order to adapt or, rather, to dominate the environment in which it lives. This reversal is physically represented in the figure of the cyborg, whose trajectory originated in a very specific context and crosses the path of transhumanist anthropology, generating highly significant results. The cyborgisation of the human being is simply a way to accelerate and push human evolution in directions that are already set out in the path followed up until the early twenty-first century (Warwick, 2004, 1). The transhumanist project is entrusted to biological engineering and aims to create a new human being (2.0 or 3.0), who has or could have the appearance of a cyborg. As Warwick indicates in the introduction to *I, Cyborg*, the changes he underwent basically entailed the insertion of technology into his body to make him a cyborg, part machine and part man. However, the view that "we are becoming cyborgs" (Kurzweil, 2005, 207) is even more radical than it may at first appear. It has been stated that rather than becoming cyborgs we already are cyborgs, in the sense that our nature is already imprinted with the possibility of our cyborgisation or, even more decisively, that we are "natural born cyborgs".

But what does the view that we are "natural born cyborgs" actually mean? The literature on the subject provides two different conceptions of the cyborg: the first is an organism with exogenous devices integrated into its biological body, while the second envisions the dissolution of the body and the hybridisation of its identity into the surrounding environment. That said, the cyborg of interest here is any "human–technology symbiont", that is any being with mental functions that is made up of biological and artificial parts (Clark, 2004, 3). The key feature of Clark's definition lies in the fact that, as an organism hybridised with the external world, the cyborg would adequately capture the nature of the human being. As hybridisation can be achieved in a variety of ways and forms, it cannot be considered a modern process but must be regarded as one of the fundamental characteristics of humanity (Clark, 2004, 4).

Had it not been taken up by transhumanism or by the many theories that, like transhumanism, propose and support enhancement practices, the term cyborg would have remained locked in the domain of extremely complex scientific research or as a metaphor to designate a free, independent and asexual body with no identity, as proposed by Donna Haraway, or relegated to childish invention of the worst fiction or sci-fi movies. Instead, it has become the theoretical figure in which all the dreams of physical and mental enhancement of the human being are condensed. The posthuman, transhumanism and hyperhumanism have all seen the cyborg as the figurative representation of their theories, albeit in different ways. However, the cyborg has characteristics that differ from the transhumanist view according to which it is essentially an organism hybridised with technological artefacts.

The correct interpretation of the cyborg revolves around two ideas: its nature as a self-regulating being that always tries to achieve homeostasis with the surrounding environment and the automaticity of this form of self-regulation. The view of the *posthuman* being envisioned and sought by transhumanism as a cyborg, or rather as a "homeostatic system functioning unconsciously", seems excessive and perhaps mistaken. On the one hand, the *posthuman* being would be granted characteristics that transhumanism does not take into account and, on the other, the transhumanist *posthuman* would be an organism that *seeks* equilibrium with its environment when the very notion of equilibrium not only appears completely alien to the transhumanist imagination and theory, but represents precisely the state that transhumanism seeks to destabilise. The idea of this new human being that transhumanism embraces is much more simply a man with enhanced performance levels: more like the

Six Million Dollar Man than a man adapted to an alien environment. Consequently, it has been suggested that the *posthuman* being should not be regarded as a cyborg and even less as a holorg, as represented by the T-1000 played by Robert Patrick (Hess, 1995) in *Terminator 2: Judgment Day*, but as an Anthrobot, a synthesis of a human being and a robot, as proposed by Natasha Vita-More. The Anthrobot would simply be a coalescence of technologies capable of vicariously handling and upgrading all human functions, mental products contained in a nanotech storage system and a general control system, a mind, independent of a physical and biological medium (O'Connell, 2017, 36).

However, here the term cyborg will continue to be used to define the transhumanist human 2.0 or 3.0, not only for the issue of convenience but also because the analysis of the cyborg will highlight the characteristic reversal of the relationship between humanity and the environment brought about by transhumanism.

The origin of the term cyborg says a great deal about the meaning that this image was intended to convey because, as is well known, it appeared in the framework of research conducted by NASA in a 1960 article entitled "Cyborgs and Space" by Manfred Clynes and Nathan S. Kline, in which the authors investigate how to survive in an unnatural environment. Space travel is a highly complex scientific challenge, which requires answers as to how the biological structure of human beings must be modified in order to allow them to survive in an environment that, for the first time in their history, is not the one in which they were born. From the scientific perspective, the problem lies in the fact that crossing these boundaries produces entropy, and it is therefore necessary to rethink the human body by modifying it in such a way as to enable it to reach and remain in a state of homeostasis with an environment that is ontologically foreign to it. However, Hannah Arendt points out that this perspective is both unheard of and reductive. Unheard of because "nobody in the history of mankind has ever conceived of the Earth as a prison for men's bodies" (Arendt, 1998, 2). Reductive because it may well be true that "the human artifice of the world separates human existence from all mere animal environments, but life itself is outside this artificial world, and through life man remains related to all other living organisms" (ibid.). The cyborg is the exact definition of this unheard of and reduced being, who appears as the descendant and heir of humanity, able to survive in conditions far beyond the natural environment of which human beings are properly part.

The reasoning Clynes and Kline follow to explain the path leading to the conception of the cyborg is crystal clear: all living organisms have had to solve a series of problems related to survival by adopting solutions that constitute a compromise the organism has managed to reach with the environment in which it is located. Clynes and Kline observe that nature has succeeded brilliantly in solving the problem of breathing in different ways, by generating fish, plants and birds. The solutions to the breathing problem that all these beings share are linked to a specific environment to which they are rigorously restricted (Clynes & Kline, 1960). Natural inventiveness has indissolubly bound a living being to a specific environment: if a fish sought to live outside the water it would be unable to do so as it currently is. Clynes and Kline assume that this rigidity can be softened by taking over nature's ability to shape matter. In other words, a particularly intelligent fish with specific skills in biochemistry and physiology, up to speed on the latest progress in cybernetics and even a bit of an engineer—in short, a synthesis of Clynes and Kline with gills—would know how to breathe out of water (Clynes & Kline, 1960).

Metaphors aside, a man who sought to live in an environment where temperature, atmospheric pressure, gravity and air composition were different from those in the environment in which he was born, although not naturally suited to it but having the necessary skills to solve the problems it posed, would immediately adapt to the new living conditions in the only way he could: by transforming. Clynes and Kline not only explore the full range of devices and organs that need to be integrated and incorporated for this purpose, but they also observe that these need to function in the same way as biological homeostatic systems, that is independently of the will of the person integrating them. Otherwise, the man who had succeeded in escaping the constraints of that context but was forced to constantly reset his artificial prostheses would become a slave to the machines that made him suitable for the environment in which he finds himself (Clynes & Kline, 1960).

It is therefore with good reason that in order to define this human being as being adaptable to an unnatural environment, Clynes and Kline propose the term *cyborg*, cybernetic organism, which merely indicates a biological organism taken from its ecological niche and subjected to appropriate modifications that enable it to autonomously and involuntarily establish an equilibrium representing the essential conditions for its survival (Clynes & Kline, 1960). Originally, the cyborg was perceived as an organism in conflict with the environment in which it finds itself and with which it has to re-establish a harmonious relationship in order to avoid its end, its death. Clynes and Kline must also have great foresight, because they are well aware that the solutions to the problems posed by humanity's survival in space will have repercussions not only on scientific progress but also on human nature itself (Clynes & Kline, 1960).

Two conclusions can be drawn for the time being. First, the cyborg conveys a view that, although implicit, is fundamental to all transhumanist reflections on the *posthuman* being: fitness is the result of a modification or direct intervention in the living being and not in its surrounding environment. The moral and political implications of this reversal and of shifting the adaptive pressure onto humanity are important but, in order to be fully appreciated, it is essential to consider the initial context in which the cyborg was conceived. This will establish a metaphorical background against which its true purpose stands out, and the latter is nothing short of the denial of any moral, economic, social and political improvement of the context of life. In short, the cyborg implicitly conveys the idea that the conception of politics adopted up to now (a concerted action to find solutions that will make the world in which people live habitable) has to be abandoned in favour of one that sees it as a means for the more or less equitable distribution of enhancing resources, which are by definition individual.

Second, Clynes and Kline actually deal with the borderline case of the problems posed by an unnatural environment, such as deep space, to explain how a biological organism can survive outside its habitat. Any biological organism is in an unstable balance with its environment, although this instability is not caused primarily by the environment but is innate to the biology of the organism that at some point loses its ability to maintain an equilibrium and loses its homeostatic functions, regardless of the environment in which it is located. Environmental conditions certainly have an impact on individual homeostasis: for example, an individual deprived of adequate nutrition, clearly will no longer be able to keep their body functioning and will waste away under the pressure of environmental factors. However, environmental conditions can only accelerate or slow down the dissolution of the organism; they cannot be the direct cause. Biological organisms are intrinsically destined to waste away, which is what would happen if it did not have homeostatic mechanisms (although their effectiveness declines over time). It can therefore be deduced that the aim of a human becoming a cyborg is immortality. If it is true that the incorporation of exogenous devices serves to continually adapt the human being to constantly changing external conditions, then there is no obvious reason why this adaptation should come to an end. In other words, there is no reason why there should not be devices that oppose the tendency to waste away caused by the natural environment and the biology of the organism itself. Every vital function can be substituted by devices in such a way as to make it indefinitely resistant to any internal or external pressure. Cyborgisation is the dream of humanity that wants to become immortal.

5.7 Legitimacy of Becoming a Cyborg

Given that enhancement is a constant practice of humanity, we will try to understand whether all forms of enhancement (whether using non-traditional biomedical technologies or traditional and well-established means) produce the same results and have the same effects overall. In other words, is it possible to differentiate between the different forms of enhancement that are put into practice? Can training based on physical exercise be considered an enhancement in the same way and with the same implications as the assumption of EPO by athletes seeking performance at the limit of human potential? Can psychoanalytic therapy intervening indirectly in the neuronal structure be regarded as similar to the taking of psychotropic drugs aimed at mood enhancement? The aim is not to establish whether it is better to enhance or not to enhance, as nature has already taken on this task and chosen the former, but to investigate enhancement procedures (Coeckelbergh, 2013b) and point out their possible consequences. This will certainly solve one problem, but it will also raise a far more serious one which is foremost among the concerns of transhumanism's opponents: if it were established that all enhancements are natural, it would mean that they are all legitimate. These brief comments clearly indicate that the key issue that transhumanism raises is not so much the legitimacy of physical modification as its limits and means. While the idea that man has always tried to find ways to improve himself is acceptable, transhumanism offers a rather problematic specification of it in that it proposes to push or exceed the biological limits of individuals through the use of technology (More, 2013a, 4). Although there is ultimately nothing shocking in the fact that transhumanists see humanity as imperfect but perfectible because it is a phase in the evolution of life on the planet, which has a long-standing history and an equally long future (More, 2013a, 4), the problem appears rather to arise from the fact that this aim (although perfection can be considered an aim in itself) is pursued through direct intervention on the human body and the biological processes that keep it functioning.

What is at stake here is the possibility of establishing a continuity between the different ways in which human beings have been modified and enhanced over the course of evolution. It is not a choice between modifying or not modifying, and it will not even be about establishing how far to modify an individual but to determine whether the forms of the human being's constant direct/indirect, implicit/explicit, voluntary/involuntary intervention on itself are similar, while accepting the idea that people are "natural-born cyborgs" (Clark, 2004). We may well be natural-born cyborgs, but we will have to find out whether the types of technology humanity uses to get the best out of itself make it possible to establish a continuity in the evolution of the species, or if they have been so different that this supposed continuity is only a veneer concealing a substantial discontinuity and difference between the diverse hybridisations of humanity and technology. In short, the question is whether the cyborg is an undifferentiated category or whether it has internal differences such as to render it unusable in capturing the essence of the human being.

As stated above, the cyborg has always been an organism that is somehow dissolved into the environment, hybridised with every tool which then becomes its prosthesis. If this is true, then it is only because of an erroneous form of biological reductionism that we think of the identity of people as being constituted by the products of organs inside the body. In actual fact, and this has always been the case, we are hybrid beings, not isolated and not isolable from interactions with the surrounding environment with which we form a single totality. The idea that there are identities defined and delimited by the biological body is merely the reinforcement of a prejudice of Western culture. The main consequence of this prejudice is not so much the belief in the existence of a soul, a *res cogitans*, so to speak, but the belief in the existence of a brain, a "mental machine" contained exclusively in a skull or more precisely in a biological body (Clark, 2004, 4). However complex and important the brain may be, it might have no privilege and no cognitive capacity that can be isolated, since it carries out all its operations in collaboration with elements that are external to the "skin-bag" that houses it. The thesis defended by Clark is that mental acts are the product of a device composed of biological and artificial elements inside and outside the body (Clark, 2004, 24), that is the matrix that creates and develops all the forms of hybridisation that give life to cyborgs all around the natural and technological world. It would therefore be wrong to think that the mind is the product of an identifiable portion of the world precisely because it is located inside our skulls. In actual fact, our mental processes have always been the result of a collaboration between brain, body and technology. According to Clark and Chalmers (who collaborated on an initial version of the anthropological theory that sees the cyborg as the most appropriate representation of human nature), the relationship between the world and the mind is a form of fusion that becomes "active externalism", in which the elements belonging to the environment where the subject is located play a key productive role (Clark & Chalmers, 1998, 7). In other words, the mind is structurally and naturally connected to the world around it. As well as mental states like experiences, which remain internally determined, there are other states produced primarily by external modules such as beliefs regarding reality (Clark & Chalmers, 1998, 12).

The way in which technology and the outside world cooperate with the mind is presented by means of a rather trivial example, which is of considerable interest here however, because it provides a theoretical and moral justification for any kind of modification or enhancement of the human body. The above authors exemplify this cooperation, and thus the equivalence of biological organs and artificial means, of the internal and the external, by analysing the way that two people use biological internal memory or external memory in the form of a notebook.

On the one hand, there is Inga who, having to return to MOMA in New York, consults her internal memory to know where it is. The internal memory, having generated the belief that it is on 53rd Street, tells her to walk along the street that separates her from the Museum. According to Clark and Chalmers, the memory of the museum lies in her memory, waiting to be activated; once she has access to it, she can cover the mental and physical distance separating her from the Museum. On the other hand, there is Otto, who suffers from Alzheimer's disease and has no immediate and direct access to his memories. So, in order to generate the belief that the Museum is located on 53rd Street, he needs to consult a sort of external memory consisting of his notebook on which he had written the exact address.

As Clark and Chalmers write, in the final analysis his notebook performs a similar function to that of biological memory. Consequently, there is an analogy between the ways in which belief in the Museum's location is generated for Inga and for Otto. Since both Otto and Inga used the information contained in two different media in the same way, the notebook for the former, memory for the latter, it is possible to state that there is a functional analogy between notebook and memory. Clark and Chalmers come to the conclusion that, in terms of producing cognitive acts, biological memory and external memory work in the same way and can be integrated into cognitive processes with the same exact consequences and problems. Although in reality things are more complex than Clark and Chalmers's analysis suggests, because the degree of cooperation in cognitive production is certainly different for internal and external memory, the fact remains that both these media cooperate in the production of cognitive processes and can therefore be considered as equivalent, at least from this point of view. While these two mnemonic media are similar in terms of access and reliability, their positioning (inside or outside the body) no longer seems relevant. If the ease of access to internal and external mnemonic media is identical, if the resources they contain are readily available and if they contribute to the production of beliefs in a similar way, then there seems to be no significant difference between carrying the medium *in one's pocket* and carrying it *in one's head*. "If the relevant capacities are generally there when they are required," then the conditions making it possible to speak of an equivalent cooperation between internal organs and external means seem to be fulfilled and "this is coupling enough" (Clark & Chalmers, 1998, 11).

Extending this reasoning to technological means that human beings have always used, the hypothesis that humanity has always delegated certain functions by using means external to its body gains legitimacy. Moreover, as a cyborg was defined above as a biological organism that acts by also using technological prostheses and artificial tools to which it is coupled, it seems justifiable to say that we are "natural born cyborgs". However, the aspect of Clark and Chalmers's thesis that is of interest here is that it has been extensively used by Neil Levy to morally justify the manipulation of biological organs in order to enhance them.

According to Levy, it is absolutely true that the mind is not limited to the productions of the brain but spills over into the world, that it is extended and extendible to any external element capable of assisting it in its productions, as proposed by Clark and Chalmers (Levy, 2007, 29). However, the consequences Levy draws from Clark and Chalmers's proposal go far beyond the original, central problems that led to the definition of the extended mind thesis (EMT). In fact, Levy uses the EMT not to provide a more adequate account of the production of mental acts, but to offer arguments in favour of enhancement and to morally legitimise any kind of modification and manipulation of mental processes.

Levy starts from two trivial and sharable observations. First, technological progress poses no particular moral problems: no one would dream of considering the increase in a PC's computing power or the upgrading of any technological device as immoral. Science merely increases our knowledge of natural processes in expectation of this enhancement and everyone is delighted about it. Second, if a patient with Alzheimer's disease who uses a notebook to remember things, replaced it with a particularly powerful tablet device, or if a person whose leg has been amputated could use a prosthesis that would allow them not only to walk normally but also to run, not only would nobody consider it immoral, but we would all be delighted about the undeniable improvement in this person's daily life. In short, any use of tools or technology that helps people in everyday life is not only legitimate but welcomed with pleasure and enthusiasm. This, combined with the conclusion reached in Clark and Chalmers's EMT that external and internal memory cooperate in the performance of cognitive acts, albeit to different degrees, allows a coherent consequence to be drawn.

If improvements in internal memory or any other organ or function that contributes to the production of cognitive acts can be considered in the same way as improvements in external devices that cooperate in the production of these same cognitive acts, it follows that both are morally legitimate. In other words, if the body and the outside environment contribute to the production of mental states, albeit to a different degree, then alterations of one should be evaluated in the same way as alterations of the other. Levy proposes two versions of this thesis, defined as the ethical parity principle (EPP): a strong one stating that "alterations of external props used for thinking are (ceteris paribus) ethically on a par with alterations of the brain"; and a weak one stipulating instead that "alterations of external props are (ceteris paribus) ethically on a par with alterations of the brain, to the precise extent to which our reasons for finding alterations of the brain problematic are transferable to alterations of the environment in which it is embedded" (Levy, 2007, 61). The application of the EPP to the EMT makes it possible to affirm that since the mind is composed of internal and external modules and since interventions that improve external parts of the mind are not morally problematic, interventions that modify internal parts of the mind are not morally problematic either (ibid.).

These two principles, which are in clear agreement with transhumanist belief, ethically justify the manipulation and enhancement of human capabilities. Levy's arguments justify the moral legitimacy of direct intervention in the human body and biology, that is the type of intervention proposed by transhumanists. Critics of transhumanism see nothing problematic in indirect interventions aimed at improving human performance such as education, psychoanalysis, gymnastics and so on, but they consider interventions of biological, pharmacological or prosthetic manipulation for the same ends as highly problematic, not to say illegitimate. But, says Levy, if cognitive acts can be performed by means outside and inside the body and direct intervention in external means (which might be a simple upgrade or a replacement of components) is completely legitimate, why should the same type of intervention in internal organs be problematic? In other words, if a man is a cyborg, and it therefore seems totally legitimate to replace the prosthesis of an amputated arm with one made of more suitable materials and fitted with a technology that allows greater freedom of movement, why should it not be possible to intervene in muscle fibres to achieve the same ends? Levy's thesis allows us to state that if it is morally legitimate to replace an external memory with a larger and more powerful one in order to improve an individual's intellectual performance, then direct intervention on brain matter, for example transcranial magnetic stimulation, for the same purposes is equally justifiable.

Let us summarise the important points raised so far. First, the idea that transhumanism seeks to transform the human species or even aims to eliminate it and replace it with a different species through direct manipulation and alteration of its biology does not reflect the intentions of transhumanists. The resulting individual or species, whether a cyborg or an Anthrobot, would be no different from *Homo sapiens*: it is not a contemporary creation as the terms used to define it might suggest; on the contrary, they define the very essence of the human being with an accuracy never previously reached. Not only does the cyborg encapsulate what man has always been far better than many other images, but within this category there are no differences between different forms of cyborg: those who simply use glasses to restore their sight, those who have chips implanted in their body or those who allow the alteration of their genetic code are hybridised with the artificial in different ways, but they remain basically within the same anthropological category.

Levy's ethical parity principle provides further justification for the equivalence between the various enhancement practices developed by humanity throughout its history. For Levy, not only is there an anthropological continuity between *Homo sapiens* and the cyborg, but the cyborg is, indeed, *Homo sapiens* finally revealed in its true nature. However, this anthropological continuity is based on an instrumental continuity. It is of little importance what technologies are used to enhance individuals: since they are always and in any case technological artefacts, there is no discontinuity of any kind. Thus, transhumanism is denied any right to be regarded as an innovation for two reasons: first, because the cyborg to which transhumanism aspires is not a figure that has never been seen before, and second, because the proposed means are no different from those used so far to enhance the species and the individual.

5.8 The Morality of Technology

Among the many objections to which Levy feels obliged to respond, there is one that is particularly worth analysing: the idea that the alteration of mental states—which can be achieved through manipulation of the brain and, more generally, of all matter contributing to these states—presupposes a mechanisation of the self that it would help to consolidate. Discussion of this topic will serve a precise aim: to show that the technologies used to enhance humanity can be differentiated not only by the aim they seek to achieve but also by the effects they produce, most of which the transhumanists and supporters of human enhancement blissfully ignore.

Supporters of the traditional means of altering mental states—psychoanalysts, physiotherapists, theologians and so on—argue that we are not machines, or at least not only and not primarily, but living creatures interacting in singular ways with external stresses and stimuli. Hence, treating mental states as objects would mean considering ourselves as things (Levy, 2007, 79). And this is exactly what we do when we use chemicals or other means that intervene directly on the central nervous system. A pharmacological transformation of the brain is based on the idea that the individual is a machine whose dysfunctions must be corrected, for example by limiting the impact of emotions in responses to external stimuli. But, Levy says, even traditional means of altering mental states treat the body like a machine. After all, what is the difference between two people trying to cure their depression, one by taking antidepressants and the other by exercising? Why do we suspect that the administration of antidepressants presupposes a mechanical conception of the body while gymnastics does not? Why, Levy asks, do we question some means of altering the mind and not others (Levy, 2007, 113)? In his opinion, these are two equally legitimate means because they have the same end, that is healing from a depressive state, and both act on the body as if it were a machine to alter and improve its mechanisms. At first sight, Levy would seem to be right: the implicit assumption of both therapies is a certain mechanistic conception of the human body. However, things are more complex than Levy seems to think, as we have already seen. Leaving aside for now all the perplexities and doubts that arise from a mechanistic conception of the human body, let us analyse Levy's reasoning, which will provide further arguments to reject it.

After rejecting the critique that objects to the supporters of enhancement having a reductionist conception of the body, Levy observes that, on closer inspection, there is indeed a difference between these two means (exercise and antidepressants) but it lies simply in their differing efficacy, as exercise is slower and its outcome more uncertain than the beneficial effects produced by antidepressants. Moreover, Levy continues, the advantage of antidepressants is that they allow the patient to regain control over their behaviour, recovering from a condition that they recognised as pathological but were unable to change in any other way. Subjects who have recognised their pathology and also their inability to modify reactions or behaviours, shows an identical conception of themselves, their worth, their subjectivity and the same form of respect for their personality, whether they intend to use means of direct intervention, such as psychotropic drugs, or whether they propose to recover from this pathology through psychotherapy, which is simply indirect intervention.

Once again, however, things are not that simple. To affirm that the only difference between the means used to produce a certain type of result lies in their greater or lesser efficacy is to adhere acritically to the concept of technology neutrality which, although generally accepted by common sense, does present some problematic aspects. What transhumanists and the unconditional supporters of enhancement generally do not take into consideration, or deliberately leave aside, is the possibility that technological means have an intrinsic normative effect. In the light of much research that highlights the moral status of technological artefacts, it appears difficult to passively accept the concept of technological indifference and thus the principle of technology neutrality. And if the idea of axiological and normative neutrality of technological artefacts were to be abandoned, it would become difficult to accept the idea of a continuity of enhancing practices based on the substantial neutrality or indifference of the means and tools used for this purpose.

At this point, a further step is required to point out if there are any moral implications resulting from the use of different technologies. Two aspects of the debate on the moral status of technological artefacts are particularly significant for this problem. The first concerns the moral agency of technology, that is whether technological artefacts can be considered independent of the intentions of their inventors, designers or users in the production of moral standards or whether, instead, moral standards are produced by a collaboration between humans and artefacts. In this case it will be necessary to ask how the roles within this possible cooperative network are distributed. The second concerns the possibility that artefacts, regardless of the agency role granted to them, independently or in collaboration with individuals, only produce intentional, that is desired, effects or whether their moral agency extends beyond intentional and desired forms of normative production and produces effects unintentionally.

The first point is concerned more with the extension of moral agency to technological artefacts than with acknowledging that the artefacts have an agency capability, that is the ability to incorporate values and norms. The crucial point is to establish whether artefacts are vehicles of intricate values or if they are only used as simple tools to implement behaviours and are thus devoid of any autonomous moral agency. One group of scholars sees technological objects purely as tools that, although substantially neutral, have the ability to incorporate the values of those who invent, design and use them. This does not mean simply attributing neutrality to technology but, rather, considering objects as passive vectors of values and norms. As a result, their normative function cannot simply be ignored, and nor should it be overestimated. Supporters of this view thus attribute a normative function to technology, which is somewhat limited to the intentions of its users. For example, the Paris metro has seen its decor change over the years: long benches with a flat surface have been replaced or supplemented with individual seats or parallel iron bars in the shape of an inverted U set at different heights to allow people to sit on the lower bar and lean back against the higher bar. The intention behind this change was to make it impossible for homeless people to lie down to sleep in the subway, an aim that has been achieved. Clearly, the normativity these bars achieve is not intrinsic to their shape or their solidity but is produced by the intentions of those who set themselves a political and social problem that they solved by adopting these technological means. However, in and of themselves, not only are the U-shaped iron bars not intended to produce or prevent a behaviour, but they are also morally neutral. It can therefore be said that the technological artefact is not a moral agent but a tool of human moral agency in which it participates passively (cf. Winner, 1989).

A second position that represents a strong version of the alleged normativity of technology gives it an intrinsic ability to produce norms, irrespective of the intentions of their inventors and users. Technology may be morally autonomous and productive, even if several authors who support this view disagree on the actual productive role of technological artefacts. It has been argued that specific mental acts, such as intentions, can be independent of individuals and may also be produced by inanimate objects. This seems to be Latour's view when he argues that artefacts are part of a network of actants that is also composed of humans.

Finally, Borgmann recognises that artefacts have the ability to produce different kinds of norms and values (moral, political, economic and existential) but he seems to be seeking a compromise between the first two positions, thus defining a particularly relevant theory on this problem. According to Borgmann, we build technological artefacts that implement values which in turn shape our choices about other artefacts, thus building a network of reciprocal influences (Mitcham, 2014, 23).

Borgmann's position seems to better represent the relationship between human agents and tools, not least because it focuses on the definition of a differential element of biotechnologies that consists of the effect they have on the people who use them. From this perspective, Philip Brey's reflections also seem particularly useful because they focus on the different way artefacts convey values, on a microphysics of interactions, thereby sidestepping the need for a massive consideration of artefacts as a "category". For Brey, the different entities play a major role in ethical structures. These entities, which can be both human and non-human agents, are all *moral* factors in the sense that within these structures they exert a "moral influence". Aside from the fact that this influence may differ in kind (positive or negative, accidental or intentional), the most fruitful aspect of Brey's reflection and, in fact, of the whole debate on the moral agency of technological artefacts, is that they are recognised as being able to act on human behaviour regardless of their mere instrumentality. In other words, the classic thesis that the technological tool is neutral but is exploited as a promoter and instantiator of human values is abandoned in favour of a more careful analysis of the forms of autonomous agency of artefacts (Brey, 2014, 137).

Secondly, embedded in all the theories on the moral status of artefacts is the idea that we must also consider the consequences that go well beyond those imagined by the inventors and designers of technological objects. Artefacts unintentionally convey values and implement behaviours, that is they seem to be able to modify the lives of people, the world in which they live and the relationships between them all in an uncontrollable and autonomous way. The type of technology with which we come into contact when we act in the world produces in itself a change in our assessment of the surrounding world. Moreover, it serves as a screen that we put between us and the world, which in turn produces a series of changes in assessments, perceptions and values. The idea is that technology represents a medium between the subject and the world and contributes to the way in which the former sees and interrelates with the latter (Verbeek, 2011).

At least two conclusions can be drawn from the above. First, the technologies used are not indifferent to the results or to the values they implement. It is clear that the values implemented by enhancement technologies are efficiency, performance and productivity, but above all fitness. The first three are compatible and consistent with the neoliberal vision of Western society, but the fourth seems to be no less than a constituent element. Enhancement technologies enable the elimination of differences in the guise of a search for identity, which in most cases (at least for transhumanists) corresponds to a greater ability in social uniformity, as is evident especially in the use of cognitive and psychological enhancement.

Second, we can confirm the view that transhumanism is unlikely to achieve its aims, which are essentially increased control over oneself and an extension of one's freedom. In fact, the technological mediation that should make it possible to achieve these aims introduces an element of heteronomy, which does not seem to be entirely controllable by individuals: using drugs to enhance one's mood, in whatever way one wishes to interpret this concept, means allowing the drug in question to take control of our mood, becoming dependent on that drug and, above all, relinquishing the possibility of developing an autonomous strategy.

Moreover, as we have seen, the effects of technology are largely unintentional: they are completely beyond the control of both their inventors and their users. It therefore seems contradictory to want to increase control over oneself and one's freedom with artefacts to which one can become dependent and which are absolutely beyond the control of the individual who uses them.

The analysis of the moral agency of enhancement tools therefore leads to the same conclusions we arrived at in the previous sections: not only does the mechanisation of the body lead to irremediable aporia, but it is the idea that one can take control of one's body and one's self through the use of technological artefacts—whatever they may be—that is fundamentally wrong, leading, among other things, to conclusions in contradiction with the effects sought.

References

Agar, N. (2014). *Truly human enhancement: A philosophical defense of limits*. MIT Press.

Arendt, H. (1998). *The human condition* (2nd ed.). University of Chicago Press.

Bostrom, N. (2009b). Dignity and enhancement. In B. Lanigan (Ed.), *Human dignity and bioethics* (pp. 123–144). Nova Science Pub Inc.

Bostrom, N., & Sandberg, A. (2009). The wisdom of nature: An evolutionary heuristic for human enhancement. In J. Savalescu & N. Bostrom (Eds.), *Human enhancement* (pp. 375–416). Oxford University Press.

Brey, P. (2014). From moral agents to moral factors: The structural ethics approach. In P. Kroes & P.-P. Verbeek (Eds.), *The moral status of technical artefacts* (pp. 125–147). Springer.

Buchanan, A. (2011b). *Beyond humanity?* Oxford University Press.

Cabrera, L. Y. (2015). *Rethinking human enhancement: Social enhancement and emergent technologies*. Palgrave Macmillan.

Canguilhem, G. (1989). Machine et organisme. In *La connaissance de la vie* (pp. 101–127). Vrin.

Canguilhem, G. (1991). *The normal and the pathological*. Zone Books.

Charis Hall, M. (2017). *The bioethics of enhancement: Transhumanism, disability, and biopolitics*. Lexington Books.

Clark, A. (2004). *Natural-born cyborgs*. Oxford University Press.

Clark, A., & Chalmers, D. (1998). The extended mind. *Analysis, 581*, 7–19.

Clynes, M. E., & Kline, N. (1960). Cyborgs and space. *Astronautics, 13*, 27–31.

Coeckelbergh, M. (2013a). *Human Being @ Risk: Enhancement, technology, and the evaluation of vulnerability transformations*. Springer Netherlands.

Coeckelbergh, M. (2013b). Enhancement and the vulnerable body: Questioning some philosophical assumptions. In F. Lucivero & A. Vedder (Eds.), *Beyond therapy v. enhancement* (pp. 15–26). Pisa University Press.

Daniels, N. (2009). Can anyone really be talking about ethically modifying human nature? In J. Savulescu & N. Bostrom (Eds.), *Human enhancement* (pp. 25–42). Oxford University Press.

Dawkins, R. (1986). *The Blind Watchmaker: Why the evidence of evolution reveals a universe without design*. Norton & Co.

Doyle, J. (2018). *What does it mean to be human? Life, death, personhood and the transhumanist movement*. Springer.

Eilers, M., Grüber, K., & Rehmann-Sutter, C. (2014). *The human enhancement debate and disability: New bodies for a better life*. Palgrave Macmillan.

Flanigan, J., & Price, T. L. (2018). *The ethics of ability and enhancement*. Palgrave Macmillan.

Giubilini, A., & Sanyal, S. (2016). Challenging human enhancement. In S. Clarke, J. Savulescu, T. Coady, A. Giubilini, & S. Sanyal (Eds.), *The ethics of human enhancement: Understanding the debate* (pp. 1–24). Oxford University Press.
Goffette, J. (2009). Modifier les humains: Anthropotechnie versus médecine. In J.-N. Missa & L.Perbal (Eds.), *«Enhancement». Ethique et philosophie de la médecine d'amélioration* (pp. 49–63). Vrin.
Gray, C. H. (2000). *Cyborg citizen: Politics in the posthuman age*. Routledge.
Haisma, H. J. (2011). Physical enhancement. In J. Savulescu, R. ter Meulen, & G. Kahane (Eds.), *Enhancing human capacity* (pp. 259–265). Wiley-Blackwell.
Harnacke, C. (2013). Discussing enhancement and disability together. In F. Lucivero & A. Vedder (Eds.), *Beyond therapy enhancement* (pp. 69–94). Pisa University Press.
Harris, J. (2007). *Enhancing evolution*. Princeton University Press.
Hauskeller, M. (2016). *Mythologies of transhumanism*. Palgrave Macmillan.
Hess, D. (1995). On low tech cyborg. In C. H. Gray (Ed.), *The cyborg handbook* (pp. 371–378). Routledge.
Israel, G. (2004). *La macchina vivente. Contro le visioni meccanicistiche dell'uomo*. Bollati-Boringhieri.
Juengst, E. T. (1998). What does enhancement mean. In E. Parens (Ed.), *Enhancing human traits* (pp. 29–47). Georgetown University Press.
Kant, I. (1991). The contest of faculties: A renewed attempt to answer the question: 'Is the human race continually improving?'. In *Political writings* (pp. 176–190). Cambridge University Press.
Kass, L. R. (2002). *Life, liberty and the defense of dignity: The challenge for bioethics*. Encounteurs Book.
Kolnai, A. (1976). Dignity. *Philosophy, 51*, 251–271.
Kurzweil, R. (2005). *The singularity is near*. Penguin.
Lanigan, B. T. (2009). *Human dignity and bioethics*. Nova Science.
Levy, N. (2007). *Neuroethics: Challenges for the 21st century*. Cambridge University Press.
Macklin, R. (2003). Dignity is a useless concept. *British Medical Journal, 327*(7429), 1419–1420.
Macklin, R. (2006). The new conservatives in bioethics: Who are they and what do they seek? *Hastings Center Report, 361*, 34–43.
Mehlman, M. J. (2009). *The price of perfection: Individualism and society in the era of biomedical enhancement*. Johns Hopkins University Press.

Mitcham, C. (2014). Agency in humans and in artifacts: A contested discourse. In P. Kroes & P.-P. Verbeek (Eds.), *The moral status of technical artefacts* (pp. 11–30). Springer.
More, M. (2011). True transhumanism: A reply to Don Ihde. In G. R. Hansell & W. J. Grassie (Eds.), *H+ transhumanism and its critics* (pp. 62–66). Metanexus Institute.
More, M. (2013a). The philosophy of transhumanism. In M. More & N. Vita-More (Eds.), *The transhumanist reader* (pp. 3–17). Wiley-Blackwell.
More, M. (2013c). A letter to mother nature. In M. M. Max & N. Vita-More (Eds.), *The transhumanist reader* (pp. 449–450). Wiley-Blackwell.
O'Connell, M. (2017). *To be a machine: Adventures among cyborgs, utopians, hackers and the futurists*. Doubleday.
PCB. (2003). *Beyond therapy: Biotechnology and the pursuit of happiness*. President's Council on Bioethics. Harper & Collins.
Peters, T. (2003). *Playing God: Genetic determinism and human freedom*. Routledge.
Roache, R., & Clarke, S. (2009). Bioconservatism, bioliberalism, and the wisdom of reflecting on repugnance. *Monash Bioethics Review, 28*(4), 1–21.
Rubin, C. (2009). Commentary on Bostrom. In B. T. Lanigan (Ed.), *Human dignity and bioethics* (pp. 145–147). Nova Science.
Sandel, M. (2004). The case against perfection. *The Atlantic Monthly*, 1–10.
Satz, A., & Silvers, A. (2000). Disability and biotechnology. In T. Murray (Ed.), *Encyclopaedia of ethical, legal, and policy issues in biotechnology* (pp. 173–187). Wiley-Interscience.
Savulescu, J., Sandberg, A., & Kahane, G. (2011). Well-Being and enhancement. In J. Savulescu, R. ter Meulen, & G. Kahane (Eds.), *Enhancing human capacities* (pp. 3–18). Wiley-Blackwell.
Shakespeare, T. (2014). Foreword: Five thoughts about enhancement. In M. Eilers, K. Grüber, & C. Rehmann-Sutter (Eds.), *The human enhancement debate and disability: New bodies for a better life* (pp. ix–xiii). Palgrave Macmillan.
Sorgner, S. L., Jovanovic, B.-R., & Grimm, N. (2013). *Evolution and the future: Anthropology, ethics, religion*. Peter Lang.
Tamburrini, C., & Tännsjö, T. (2011). Enhanced Bodies. In J. Savulescu, R. Ter Meulen, & G. Kahane (Eds.), *Enhancing human capacity* (pp. 274–290). Wiley-Blackwell.
Verbeek, P.-P. (2011). *Moralizing technology: Understanding and designing the morality of things*. University of Chicago Press.
Warwick, K. (2004). *I, Cyborg*. University of Illinois Press.

Weckert, J. (2016). Playing God: What is the problem? In S. Clarke, J. Savulescu, T. Coady, A. Giubilini, & S. Sanyal (Eds.), *The ethics of human enhancement: Understanding the debate* (pp. 87–99). Oxford University Press.

Wiesing, U. (2008). The history of medical enhancement: From *Restitutio ad Integrum* to *Transformatio ad Optimum?*. In B. Gordijn & R. Chadwick (Eds.), *Medical enhancement and posthumanity* (pp. 9–24). Springer.

Winance, M., Marcellini, A., & de Léséleuc, E. (2015). From repair to enhancement: The use of technical aids in the field of disability. In S. Bateman, J. Gayon, S. Allouche, J. Goffette, & M. Marzano (Eds.), *Inquiring into human enhancement: Interdisciplinary and international perspectives* (pp. 119–137). Palgrave Macmillan

Winner, L. (1989). *The whale and the reactor: A search for limits in an age of high technology*. The University of Chicago Press.

6

Transhumanism and Biopolitics

6.1 Introduction

Contemporary philosophy and science have fully assimilated the idea that humanity is a product. Biology has successfully explained that our species is the result of a seamless succession of random events that have altered it to such an extent that it is impossible to identify a form and a nature specific to the human species. Cybernetics and cognitive sciences force us to engage with the autopoietic abilities of living organisms that must necessarily be regarded as being in constant transformation. Philosophy is increasingly accrediting the idea that in order to escape the sterile complications of liberal individualism, which have denoted the theoretical configuration of modernity at least since Locke, it is necessary to examine all the possible forms of hybridisation in humanity's theriomorphic and technomorphic environment. Humanism, for its part, is investigating the best way to harness the power of biotechnologies and biosciences in order to replicate its control and social selection functions which up until now have been guaranteed by literary culture.

F. P. Adorno, *The Transhumanist Movement*, https://doi.org/10.1007/978-3-030-82423-5_6

Transhumanism has inherited the sublimation of all these movements and tendencies, which it condenses into a synthesis aimed at empowering humanity and achieving immortality. Of course, there remain some grey areas in this triumphant picture, some elements that engender a certain confusion regarding the procedures and aims of human transformation, as well as its validity and theoretical legitimacy. It would be reductive to think that the answers to these questions are the responsibility of ethics, medicine and science. Restricting the scientific fields involved in this discussion to sciences like biology seems to have the implicit purpose of legitimising science and technology a priori, because it has been stated that every attempt to limit the scientific and technological expansion of modernity can only be the result of unfounded fears that hinder an overall understanding of its benefits and the importance of progress. Instead of criticising scientific progress (which is in any case futile because it is unstoppable), it would be far more intelligent to valorise its positive aspects, indicate morally correct research goals and then try to control any excesses. This antithesis is essentially a sterile contrast between equally valid theses (in that they are equally unverifiable) and ends up endorsing the supporters of enhancement and transhumanism whose views seem to be legitimised precisely by this sense of futility. It is therefore necessary to take a further step towards political considerations because the political configuration of modernity constitutes the normative framework that accepts, justifies and promotes one type of scientific research to the detriment of others. Criticism and resistance to the development of power over life cannot be conducted on the same level as the knowledge disciplines working towards that goal but must question the normative structure regulating them. Only by integrating the theoretical-political framework into the analysis of the impact of life sciences will it be possible, on the one hand, to understand that the purpose of scientific research is defined by a set of imperatives extending far beyond the scientific ones and that the boundary between science and ideology is extremely blurred and, on the other hand, to highlight the effects and involuntary modifications it engenders. This criticism of biosciences is not dictated by fear of uncontrollable developments or by the desire to foreshadow forms of scrutiny over scientific progress, but by the conviction that science is not an ontological totality that is sufficient in itself and by the awareness that it does not possess an internal and intrinsic moral legitimacy.

With crystal clarity, Heidegger highlighted the error inherent in every undue extension of biology (which is here assumed as an example of all other sciences) as well as the grounds that would justify it. If biologism is the "extension—and perhaps exaggeration and transgression of boundaries—of biological thinking beyond its own realm", which suggests an unfounded violence of thought and a confusion of knowledge, then it becomes legitimate to question its origin. For Heidegger what goes wrong in biologism, however, is not merely the transfer and unfounded extension of concepts and propositions from the field proper to living beings to that of other beings; what goes wrong already lies in the failure to recognise the metaphysical character of the propositions concerning the field, propositions by which all biology that is genuine and restricted to its field points beyond itself (Heidegger, 1991, 45). In other words, continues Heidegger, "biologism is not so much the mere boundless degeneration of biological thinking as it is total ignorance of the fact that biological thinking itself can only be grounded and decided in the metaphysical realm and can never justify itself scientifically" (ibid.). Transferred to the epistemological, moral and political level, this need to reveal the foundation of science leads to a path that starts from an analysis of the general configuration, the theory and the practices of transhumanism, and distils its meta-theoretical difficulties and epistemological assumptions to seek and analyse its political and moral foundation.

This is the view expressed by Hannah Arendt in *The Human Condition*: given the possibility of humanity to remove itself from its natural environment or profoundly modify its nature and hence escape the human condition overall, Arendt observes that the answer to the question of whether it is right to use scientific and technological knowledge in this way cannot be supplied by science. Since this question concerns the value of knowledge while science deals with facts, the answer can only be political (Arendt, 1998, 3).

Transhumanism works on assumptions which are never discussed as such and which, above all, have to do with a specific anthropological conception based on the presumed incompleteness of the individual's biological heritage, which constitutes the justification for their enhancement. In actual fact, transhumanism aims to serve as a generalised therapeutic system whose explicit aim is to restore or attribute faculties capable of improving individual wellbeing through increased freedom. However, the thesis whereby man is a fundamentally inadequate entity or naturally in a state of virtual deficiency and lack, risks triggering a pathologisation of existence according to which all observed imperfections are defects to be eliminated, disabilities to be compensated for and diseases to be cured. Following this line of interpretation, attempts have been made to show that transhumanism often translates into practices that fiercely contradict its underlying assumptions, thus raising the suspicion that it is a theoretical galaxy denoted by anti-liberalism. A tendency towards anti-liberalism may be the result of a heterogenesis of aims which, although exonerating transhumanists from seeking to pursue ends that contradict their manifest intentions, is a danger that is never explicitly denied. The thesis I aim to set out is that, since contemporary politics can be defined as biopolitics (i.e. a form of politics which elects biological life as its main object and aims to control, manage and govern the life force of people, taken both singly and collectively, i.e. as individuals and as members of a population, *omnes et singulatim*), transhumanism is merely a tool that is more or less aware of its role within this arrangement.

If other evidence were to confirm this first impression, it might shed doubt on the view that transhumanism is the movement or compendium of theories seeking to contribute to human liberation and autonomy, as is frequently stated. However, transhumanism might suffer the same fate as liberalism: born to liberate, it has become an instrument of domination that reproduces an extension of the government of individuals under the guise of progress and individual enhancement. This chapter will explore the biopolitics embedded in the actions and beliefs of those who support the philosophy and values that make up the theoretical arsenal of transhumanism.

6.2 Biopolitics and Transhumanism

Solving this complex conundrum will entail putting our imagination to the test and identifying the epistemological and political context that made transhumanism possible. Let us therefore start with the elements that have so far appeared to be structural in transhumanism. First, transhumanism places the emphasis on the *materiality* of the human being, on its biological medium. The second element is the importance attributed to *techniques of direct intervention in* human biology. Third is the common and precise purpose that these techniques share, namely the *enhancement*, augmentation or improvement of the human body's physical performance. A fourth element, which has so far remained in the shadows, is the fact that these changes impact on politics and at the same time are made possible by a *specific configuration of modern politics*. Transhumanism is a theory which argues that *direct intervention in the human body with a view to its "enhancement" is a moral and political goal in and of itself*. As we shall see, these four elements are found in the structuring of the biopolitical system characterising modernity, which steers the government of individuals towards a specific purpose: the *production* and *management*, in the broad sense of these terms, of *the biological substance of the human body*. It is precisely the description of this system that will help to understand the problematic nature of the alleged liberalism of transhumanism.

As we have seen, transhumanism is proposed as a family of theories that are particularly attentive to the positive potential of science and in this it seems only to repropose the characteristics of liberalism in another context, which make it particularly acceptable. Transhumanism suggests increasing freedom, breaking the "biological" chains that still prevent us from doing and being what we want. This liberating push assumes the form of a reduction in submission to biological processes that take place outside the control of individuals. If we limited ourselves to taking note of this convergence, there would be no further questions to ask; and to the question of the compatibility of transhumanism with liberalism, we could only answer enthusiastically in the affirmative. At the same time, the critics of transhumanism would be nothing more than lingering and fearful conservatives, if not, worse, enemies of the human species who prefer to see it suffer unnecessarily under the domination of forces that could ultimately be controlled and dominated. This oppositional reading (conservatives vs liberals; transhumanists vs bio-luddites) appears too binary and simplifying, too superficial and fundamentally erroneous, not in the sense that it contains errors, but because it does not grasp, in my opinion, the real stakes in the transhumanist themes, represented by a politics of the body. Transhumanism, in what we can consider its two versions, the one aimed at biological enhancement and the other centred on the elimination of the body possibly produced by the development of artificial intelligence in the period of singularity, implicitly constitutes the human body as the political object par excellence. In a way never seen before, the body becomes a political body, a body subject to politics and a body as the field of application of politics. This is certainly an innovative element of the transhumanist movement: modern and contemporary political theories from Hobbes onwards have never proposed to raise the body as the preferred object of politics that would have the task of optimising, manipulating and modifying. This lack explains their relative uselessness for the purposes of an adequate understanding of transhumanism. It is only since Marx that a very partial reflection on a politics of the body has been developed, but it is more decisively in the twentieth century, following a whole series of changes in economics, medicine, biology and statistics, where transhumanism is the ultimate development, that the awareness was born that the main object of politics and therefore the field of confrontation between different political theories was the human body, or more precisely the biological processes that are implemented in human bodies. Schematising, we have gone from a form of passivity towards the body, in which the biological dynamics suffered, which is why politics did not deal with them directly, to the ability to modify them, thus making them enter the field of controllable and therefore manageable phenomena by politics. Biology has in some way been extracted from the private sector and considered a "public thing", literally consubstantial with the *res publica*, and therefore to be managed and governed. Recent political theories, for example liberalisms or utilitarianisms, have concealed these aspects of the exercise of power or, according to a more benevolent reading, have not shown great sensitivity towards them, thus depriving themselves of the power to fully understand the political dynamics, more or less consciously conveyed by transhumanism. For this reason, it seems that the use of the categories that merge the biological with the political is particularly useful: it allows us to see that underneath the binary oppositions conducted in the classical terms of political philosophy, what is at stake is a transformation of the political and moral dynamics, a shift (or integration) of their centre of interest and anchorage from the classical problems of the distribution of wealth, respect for individual freedoms, the definition of the criteria of sovereignty, and so on, to the management and governance of the biological processes of the individual body within the body of the population. This means that political configuration of modernity enabling and legitimising the biosciences and biotechnologies is centred on life. The term "biopolitics" proposed by several quarters adequately captures the specificity of contemporary politics that has chosen biological life as its main focus. Thus, starting from the assumption that contemporary politics is focused on the purpose-driven management of biological phenomena, it is evident that in order to understand whether transhumanism represents a liberation practice or instead heralds a specific form of dominion and control, it will be necessary to analyse its relationship with this general context. In other words, questioning the coherence and compatibility of transhumanism with the biopolitical system appears to be a fully legitimate perspective to adopt in order to understand its possible emancipatory capability.

It must be said that even some transhumanists have perceived this link between biopolitics and transhumanism, but their conception of biopolitics appears utterly inadequate to represent the phenomena, techniques, aims and the political configuration that has life as its object. Biopolitics is considered simply as an area within which to specify how biotechnology must be used in order to avoid its abuse. Political oversight of the empowering aspects of biotechnology should focus on the preparation of a kind of code of conduct establishing the limits and legitimacy of its use. In conclusion, the transhumanists simply lament the absence of discourse and a public policy focused on defining key concepts of our biology that can provide a moral and political regulatory framework governing the use of these technologies (More & Vita-More, 2013, 279–360; Reiner, 2013; Staman et al., 2008). Biopolitics is thought of as a domain in which contenders with different conceptions of life fight it out and come up with equally different indications of the means for using biotechnologies and the ends that should be pursued. Tritely, bioconservatives and bioliberals decline specific versions and applications of their essentially traditional theories on the questions raised by the emergence of biotechnology, thus making biopolitics a branch of politics focused on biotechnology instead of economics and education (Hughes, 2004, 53–73; Savulescu & Bostrom, 2009, 1). The perspective adopted here is much more complex, starting from the idea that there is a general contemporary political system based on the fact that life has become the object of politics in a very specific way. It is only after the concepts, functions and aims of this system have been defined and clarified that issues related to the use and abuse of biotechnology can be correctly understood.

A good starting point for these thorny problems is the abovementioned analyses by Arendt, not only because of their intrinsic importance but also because transhumanism seems perfectly consistent with the picture of modernity sketched by the German philosopher, even though it cannot be considered her target for obvious chronological reasons. By adequately introducing the definition of biopolitics, this theoretical line will help achieve a deeper understanding of the transhumanist project.

Within the context of a vast genealogical reconstruction of the condition of the modern human beings, Arendt pinpoints the theoretical elements that recode politics in modernity while stressing its dependence on essentially economic and social contexts and practices, which were originally excluded but are now at the origin of the current structural crisis of modernity. This reconfiguration was made possible by the insertion of life into the political context, the secondary but no less salient effect of which was to put labour at the centre of the concerns of the individual for the first time in history. Originally intended simply as a means to satisfy the needs of life, labour became the main focus of politics, supplanting other forms of activity such as work or action. Unlike the latter activities, labour guarantees the reproduction of the individual, the multiplication of the species and the development of manufacturing to supply the goods needed to ensure both, but it is also a process of transforming nature, the life cycle of which subjugates the individual. Reduced to the state of *animal laborans*, forced by the necessities of the life cycle and distanced from the exercise of freedom, the human being is under the yoke of the needs linked to the reproduction of its existence, like all other animals. This dependence makes it impossible for it to set goals for its existence other than reproduction and the enhancement of life.

Arendt's diagnosis is merciless: the introduction of life into the framework of politics places labour at the centre of everything and, as a result, gradually replaces the political, that is the sphere of production and reflection of freedom, with the social. The introduction of economic aspects into the more clearly political sphere, that is the hybridisation of two disciplinary areas and two organisations of existence, ontologically different in origin, inevitably leads to the crisis of politics which is reduced to political economics, a veritable oxymoron for the ancient Greeks who believed that anything linked to the needs of existence was not pertinent to politics but somehow created the conditions for it and thus had to be excluded (Arendt, 1998, 29).

Two lessons can be drawn from this to ensure a proper understanding of transhumanism. The first concerns the politics of transhumanism which, in light of the union of economics and politics celebrated with terms, metaphors, images and concepts derived directly from biology, cannot be considered simply a sort of morphological bioliberalism, but must be framed within the economic and social context supporting it. This system is only apparently aimed at liberating people from constraints that burden and hinder the full achievement of their individual potential but which, in actual fact, uses this trend to propose a very specific governance of individual behaviour. The second lesson is a direct consequence of this consideration and states that in transhumanism the search for autonomy and freedom is a smokescreen concealing the construction of a political, moral and economic order that is just as oppressive as the one it intends to bring down.

6.3 Biopolitics as Life Economics

As Arendt's analyses point out, modernity is seeing a reconfiguration of the exercise of power that is intervening more and more profoundly in the management of individuals' life cycles. The shift in the perspective of modernity away from the ancient tradition is probably summarised even more effectively by Foucault who writes that "for millennia, man remained what he was for Aristotle: a living animal with the additional capacity for a political existence; modern man is an animal whose politics places his existence as a living being in question" (Foucault, 1998, 143). For Foucault, from Aristotle until a fairly recent period, politics was a practice and a problem to be faced after having solved all the problems inherent in the biological structure of the living, food, sustenance, shelter, after having solved the primary or natural needs of the living human being as a living being. Politics was by definition extraneous to bodily and biological dynamics which were relegated to another, less noble and narrower sphere. It was the family economy (basically a redundant expression insofar as economy and family have the same root, *oikos*) that had to deal with it with means that had nothing political at all. The change observed in the contemporary era, due to the fact that it is now considered the task of politics to satisfy these primary needs, is producing a disastrous collapse (at least from the point of view of a certain conception of politics) of the political over the biological. This difference which the Greeks held firmly to was founded on a double conception of life.

Although there is no unanimous view on this matter, in all likelihood it can be assumed that the Greeks distinguished *zoe* (organic existence, the simple fact of living and subsisting as a biological organism) from *bios* (understood as a characterised life, a life that has a moral and political dimension). The term *zoe*, indicated life as a raw biological fact represented by processes that enabled the subsistence and reproduction of the organism, whereas the term *bios* indicated the way in which the subject tried to give a form, a direction and a meaning to their biological life, to their mere existence.

One of the essential ways of exercising biopower is to simultaneously establish and dissimulate a hierarchy of these two notions: passing through *zoe*, biopower governs people *omnes et singulatim*, and this governance is in actual fact a rewriting of their individual, shaped lives, that is their *bios*. According to Foucault, contemporary biopolitics reduces the distance between *zoe* and *bios*—these two notions of life that the Greeks scrupulously kept apart—so that the first effect, the first instrument of governance used in the context of contemporary politics, can be regarded as the very desire to confuse these two aspects in order to better govern the life of individuals through the control and management of people's biological needs. Biopolitics is thus composed of two moments: the governance and manipulation of *zoe*, which is the framework within which the formation of individual life (*bios*) must necessarily take place. More precisely, because *bios* is constituted in relation to *zoe*, it appears to be completely managed and controlled by external forces and is therefore under the dominion of external forces.

However, before focusing on the consequences of this interpretation of modernity—essentially characterised by biopolitics and the exercise of a biopower (a power over life) shared and developed by Foucault in directions that reveal the theoretical foundations of transhumanism—it is necessary to answer a question that may decree the futility of the analyses that follow. One of the major objections to this interpretation of the political system of modernity proposed by Arendt and, above all, by Foucault, concerns the very fact that the term biopolitics may conceal a new and specific feature of modernity. To maintain that biopolitics is specific to modernity is to imagine that its object was different prior to modernity. But what has power constantly occupied itself with if not the life of individuals? What could it concern itself with if not the government of people? How can a technique of power that does not deal with the lives of citizens and people in general be conceived? In conclusion, it is claimed, biopower cannot be specific because *bios* has always been an essential and universal element of politics and the matrix of all historic forms of power. If this were the case, it would no longer be possible to claim that transhumanism belongs to the biopolitical framework and thus it would likewise not be possible to maintain the view that transhumanism will make it possible to achieve goals that are in contradiction with what they themselves claim to want.

However, supporters of this objection do not take into account the differential elements between regimes of power which make it possible to see the modern specificity of biopolitics beneath an apparent continuity. The means, techniques and ends according to which life is "captured" by politics change through history and justify the fact that we can speak of a specific exercise of power that becomes a specifically characterised biopower in modernity. While Arendt claims the fracture concerns a change in the goals of politics beleaguered by scientific and economic changes leading to the progressive expansion of labour to the detriment of other forms of human action, Foucault's analysis focuses on a change of purpose, on a different type of relationship between power and its object. Modernity is certainly a period of anthropological modification that abandons the Aristotelian conception, but it is also a period in which the action and purpose of power substantially change. This change can be summarised in the formula identifying the passage from the power of the sovereign ("take life or let live") to the power of modern governance ("foster life or disallow it to the point of death"). Unlike the power of the premodern sovereign, modern power is exercised in the form of "to incite, reinforce, control, monitor, optimise, and organise the forces under it" because it is a "power bent on generating forces, making them grow, and ordering them, rather than one dedicated to impeding them, making them submit, or destroying them" (Foucault, 1998, 136).

This shift from sovereignty to biopolitics, to use Foucault's own terms, also produces a change in the way power is structured around the new object, with a leading role being given to economics. Up until the end of the sixteenth century, economics was fundamentally a means of quantifying the wealth of the sovereign but had no function in the actual exercising of power, whereas from this moment on, those who exercised power sought their legitimacy in economics. The prince or other sovereign would know if he was governing well and thus if his power was stable by observing that his government respected the laws of economics. This means that the only discipline able to dispense legitimacy and limit the power of rulers is now economics.

The history of the birth of biopolitics is therefore the history of the way in which a system comprising economics, biology and medicine latched on to the body in order to put it to work, maximise its production and develop its strengths. The implicit and explicit aims of biopower are to supervise and collect, control and produce, protect and develop.

It could be said that biopolitics is achieving the long-desired pacification of humanity and its animality; it marks the end of history—the era of conflicts—and opens up the epoch of the game, as Alexandre Kojève had understood. The reconciliation of humanity with its animal nature and the acceptance of its continuity with the animal kingdom achieved through biopolitics appear to mark the liberation of humanity from the obligation to work and from the violence of conflict but, on closer examination, they actually root it even more deeply in its biological substratum through all forms of manipulation and hybridisation as well as the indiscriminate and total development of its body. As Nikolas Rose suspects, the new biotechnologies are not simply new ways of hybridising the organic and the artificial, but tools at the service of biological life itself, so that "in the process the human becomes, not less biological, but *all the more* biological" (Rose, 2007, 20). In other words, it is a process that, under the guise of a liberation *from* corporeality, initiates an even more rigid imprisonment *in* corporeality.

6.4 The Economics, Morality and Politics of Enhancement

One of the arguments used in favour of enhancement practices is that they would benefit society as a whole and not only specific individuals or specific social groups. Although aware that they might accentuate social inequalities, Sandberg and Savulescu are convinced that enhancement practices, above all the cognitive enhancement they focus on, will bring great economic and social benefits (Sandberg & Savulescu, 2011, 93) provided, of course, that they are well distributed and equally well used.

The primary benefit of enhancement is a reduction in economic losses. Sandberg and Savulescu claim that people will be more attentive, more aware and more in control of their actions, including those trivial, everyday occurrences that cost time and money, such as losing house keys or making a mistake in the evaluation of a financial investment. Secondly, it will benefit individuals because it will make them more attractive in the labour market and thus increase their income. Translated into clearer terms, those who can or are willing to afford enhancement will be more competitive and will therefore benefit more from the introduction and use of enhancement. However, Sandberg and Savulescu add, these individual benefits will be of advantage to society as a whole because they will reduce social costs and above all increase productivity (Sandberg & Savulescu, 2011, 97). This is a fairly astute but basically coherent application of Rawls's principles of justice, in particular the second principle whereby economic inequalities are acceptable as long as they benefit everyone. Enhancement, its supporters claim, would work in exactly the same way: they do not level inequalities but allow everyone to benefit from them, because the end results are increased productivity and the progressive accessibility of consumer goods for all, which would improve the standard of living of the entire world population, even those who have not been able to benefit directly from the enhancements.

Buchanan is even clearer on this point. He not only establishes an extrinsic and directly proportional link between enhancement and productivity but he maintains that we must stop regarding the enhancement ethic as being isolated from the context in which biotechnologies operate and finally see it as a component and a means of economic development. Buchanan even states that the history of the one is confused with the history of the other and that economic progress is the result of the self-enhancing progress of the human species (Buchanan, 2011b, 38–39). An increase in human capabilities, efficiency and productivity, *ipso facto* produces economic development and progress; reciprocally, progress in production techniques brings about an improvement in human capabilities. For example, by improving the capacity to exploit land and its produce, agriculture has made it possible to nourish people better, hence making them stronger, more effective and more long-lived, thus increasing their overall wellbeing and upgrading the quality and quantity of agricultural production and so on (Buchanan, 2011b, 44). Economics and morality are welded together in a single system which guarantees positive results through various forms of enhancement, as advocated by the transhumanists.

It is precisely this moral-economic system that was the focus of some of Foucault's reflections during his 1997–1978 course "Security, Territory, Population". Like Buchanan, Foucault recognises the existence of a direct link between economics and morality, between production and happiness, and between living and living well, identifying its root in the creation of a somewhat particular administrative body: the police. The specificity of the police as theorised from the early seventeenth century to the mid-eighteenth century lies in its object. As Foucault states, the specific object of the police is the correlation between the happiness of individuals and the increase in state power (Foucault, 2007, 422). The police do not deal with the happiness of individuals or increasing the power of the state, but they undertake to achieve these two objectives simultaneously, and hence both are factors of mutual accomplishment: if people are happy, the state will be stronger, and a stronger state will make people happier. The police aim to achieve this equation by defining a dual field of action. In order to make people happy, the police are in charge of all those activities that guarantee more than just subsistence: an adequate channelling of these resources towards the achievement of what is considered to be a good life. Situated at the intersection of the economic and social system that has to guarantee the life and survival of individuals and the will to ensure a surplus, that is an ersatz of material wellbeing, the police have the task of converting all this into state strength. Foucault stresses that the police have the aim of making the happiness and wellbeing of individuals useful to the state: the more people enjoy a pleasant and happy life, the safer and stronger the power of the sovereign and the government becomes (Foucault, 2007, 421). In actual fact, the scope for police intervention can be defined with an even better term than those used so far, ranging from happiness to amenity but all belonging to the semantic field of morality (ibid., 422). Although this term was created at the beginning of the seventeenth century, it was still quite rare in the eighteenth century: *political economics*. The creation of the police somehow made a decisive bond between the aims of economics and those of morality. If this were not enough to make its work decisive, policing was one of the tools used in seventeenth and eighteenth-century governance to ensure its preservation by maintaining the equilibrium of its internal forces. The police have essentially economic-moral aims (moral control and economic development) and serve to ensure the stability and preservation of the state as such: it is a tool for governing people to preserve the splendour of the state, which is in turn the result of the wellbeing and development of the forces of the state itself.

This is certainly not to suggest that enhancement is somehow a tool to reinforce police control over individual lives. However, establishing the genealogy of enhancement practices and assigning a seemingly imaginary origin to the aims that its supporters propose to achieve entails highlighting a fundamental element that somehow remains hidden in the reflections of Sandberg and Savulescu and Buchanan. This element is the context within which the bond between economic development and individual wellbeing has been made and a reflection has developed, not on how to help people to live but on how to help them to live well, something that can be achieved through an increased supply of material goods and improved conditions of life (in a purely biological sense); but, above all, this element is precisely the point on which to fasten enhancement practices. If we further add that the context in which these reflections are found is the definition of governance and that, from now on and increasingly, it will be on the basis of the definition of governance and of governmental rationality that economics and morality have calibrated their respective concepts, then the picture is complete.

6.5 Wellbeing Between Medicine and Politics

It could be claimed that this recourse to this analysis made by Foucault of seventeenth and eighteenth-century governance is anachronistic: after all, what relation is there between seventeenth-century police science and an ultra-technological movement projected into the twenty-first century and beyond like transhumanism? A priori, none. On the contrary, this juxtaposition is seemingly made in complete and conscious bad faith in order to artfully discredit theories and practices so innovative as to arouse unjustified fear and confusion. However, and this is my hypothesis, transhumanism and eighteenth-century police are closer than we might think because not only are their aims similar—to work for the wellbeing of individuals—but their means are not substantially different either and are essentially, though not exclusively, founded on medicine. In other words, and to be more precise, transhumanism basically claims to work for the wellbeing of individuals, for the improvement of their performances with an enhancement of the control that people can have over their body through the application of innovations in biomedicine, and, net of the various technical means, this is also the aim of the eighteenth-century police. However, we must be careful: in these two contexts, the objective cannot be considered the simple increase in the wellbeing of individuals, but it is the increase in the wellbeing of individuals through the management and governance of all the aspects of their corporeality. The benevolence of the government has a less attractive aspect which is actually an exchange: the life of individuals can become less painful, easier, people can gain happiness and wellbeing provided they submit to a series of imperatives and obligations that somehow contains them in a well-defined economic, social and economic framework. *Compelle intrare* wrote Augustine.

Of course, a simple semantic assonance between the aims that the transhumanists set themselves, the context and formulations of the nascent political economics and a reflection on police science is not sufficient to link transhumanism to eighteenth-century policing, by implicitly attributing to the former the explicitly stated aims of the latter. In order to understand how the conjunctive role of medicine can fill a temporal and theoretical gap that makes the analogy between transhumanism and the police incomprehensible, we can turn once again to Foucault and, specifically, to some essays from the 1970s (1994a, 1994b, 2001b, 2001c) in which he describes a number of changes in the role of medicine.

As mentioned above, from the late eighteenth century on and under the pressure of a significant change in the economic geography of Europe, the health and wellbeing of the population became objectives of political power. This transformation highlighted the role of medicine in managing the population as a body made up of other bodies, an organism made up of other organisms for specifically utilitarian purposes (Foucault, 2001b, 96). From this moment on, the traditional aims of medicine, that is to cure and heal, were supplemented by an additional function: to guarantee the health of individuals and the population before the onset of disease. In other words, medicine was incorporated into a new system bringing together morality, economics and politics, which progressively modified its objectives: "classical" medicine, whose purpose was to heal and to restore the biological capital of individuals, was joined by social medicine, a generalised hygienism aiming to prevent the weakening and degeneration of the social body through social measures (ibid., 100). From this moment on, society, or rather the population, became the object of a "noso-policy", a health policy that is also a "somatocracy", that is a management of bodies. Foucault describes the change of the theocratic political regime into a "somatocracy" as a general description of our society and its tendency towards a divinisation of the body in these terms: All the great theocracies, from Constantine to the mixed theocracies of the eighteenth century in Europe, were political regimes for which the salvation of souls was one of the main objectives. You could say that what has actually been brewing since the eighteenth century has arisen in the news, that is to say not a theocracy, but a "somatocracy". We live under a regime for which one of the aims of state intervention is body care, bodily health, the relationship between illness and health, etc. (Foucault, 1994a, 43). Although this statement was made in a humorous tone, it conceals an underlying thesis that Foucault summarises more clearly by stating that "the body is a biopolitical reality; medicine is a biopolitical strategy" (Foucault 2001c, 137). We can thus give a certain concreteness to the biopolitical hypothesis: it is a coupling of medicine and politics on the body of individuals and the population with a specific purpose in mind, namely their health. In fact, the overall biopolitical system is based on the necessarily consensual objective of ensuring health which justifies its coupling with the body. In other words, health has become the tool through which biopolitics has grown disproportionately, thanks partly to the instrumental and indiscriminate use of medicine, whose concepts, theory, practices and recognisable aims are profoundly modified as a result.

This appropriation of medicine for purposes other than simple therapeutics took place in several steps, as has already been discussed. But it is between the late seventeenth century and today, and especially in the mid-twentieth century, that Foucault identifies a particularly important juncture of which the Beveridge Plan is both a symptom and a representation. Although this plan was aimed at a mutualisation of social benefits and expenditures (though it actually had the objective of improving people's lives and, more specifically, guaranteeing a healthy life), on closer inspection it seems to resume and implement the aims of the eighteenth-century police (Foucault, 1994a, 40). The novelty of this plan lies in the fact that states no longer focus on the health of individuals according to their power but evidently according to the wellbeing of the citizens themselves and the benefits that this support might bring to society. The explicit introduction of health into the sphere of political concerns produced at least two important changes. The first, and perhaps the most significant, regards the scope of medicine which, now situated "on the level of life itself and its fundamental events" (ibid., 48), is pushed to go far beyond its traditional interests, to produce a movement "of indefinite medicalisation". The second, equally important but more veiled change, was the subordination of the ends, all to its own advantage, which the governance of individuals aims to achieve, and which are no longer explicitly claimed. The numerous consequences of indefinite medicalisation can be summed up in a willingness of medicine to imprint a style and a direction on society as a whole, encouraging it to reorganise itself "as a milieu of physical wellbeing, health, and optimal longevity" (Foucault, 2001b, 94), and ultimately directing and promoting one type of behaviour to the detriment of others that would be less productive in strengthening the state.

The general framework designed by Foucault clearly offers an utterly innovative perspective on transhumanism, as it frames it within a movement of change in the functions of medicine, economics and social restructuring with which it seems entirely consistent. Given the anthropology of disability that justifies its aims, transhumanism seems to be a coherent consequence of the contemporary "noso-policy" and the subsequent extension of health research. It would seemingly not be wrong to think of transhumanism as a tool in tune with the extension of biopolitical governance over the body of individuals, through contemporary somatocracy. From the context thus drawn, it becomes utterly impossible to draw a line between therapy and enhancement, a fact that transhumanism employs to justify its aims. In the previous chapters, we highlighted the somehow technological difficulties within medicine (starting from a number of succinctly summarised interventions whereby the aims of medicine changed from *restitutio ad integrum* to *transformatio ad optimum*) that make it impossible to draw a clear and precise demarcation between therapy and enhancement. Now, however, Foucault offers another more significant clue concerning the reasons that make this separation impossible: under the pressure of the processes already mentioned, medicine has progressively supplemented its therapeutic function with a tendency towards hygienism. Consequently, the therapeutic functions of the doctor were supplemented with the directive to prevent the onset of disease, which became the first and most urgent moral imperative (Foucault, 2001b, 100). Ultimately, it can be assumed that it is precisely this conception of medicine as social hygiene that facilitates the deployment of transhumanist theories.

6.6 Enhancing to Manage and Govern

In light of these considerations, the evaluation of the elements, processes and trends in medicine and biology that are so crucial to the modern world clearly cannot be left solely to morality but must also be examined by politics. In order to corroborate and defend this strong political connotation of medicine and biology, it has been necessary to move from the analysis of special and local knowledge to a higher level that incorporates it into a wider theoretical and historical perspective, namely the idea that the characteristic of modern and contemporary politics is to be biopolitics. Moreover, we have seen how medicine has been a component of governance since the seventeenth century and has even led to the theorisation of a medical police body. In this sense, it is reasonable to assume that medicine is the link between transhumanism and the techniques of governance of individuals that originated in early modernity and gradually changed up until and the mid-twentieth century when, according to Foucault, it underwent a fundamental evolution. The essential transformation that took place in this period concerns the central role that individual wellbeing has progressively acquired and that elected it as the manifest purpose of medicine, at least according to the WHO. In other words, the link between transhumanism and biopolitics, the compatibility of the former within the framework of modern politics, has been achieved by assuming individual wellbeing as a goal that politics must ensure, also and above all through the use of medicine and its entire pharmacological and therapeutic arsenal.

We can, however, take an additional step and ask ourselves what concrete changes medicine has actually undergone and which can be used to justify the view that the enhancement techniques proposed by transhumanism are equally tools that can be employed in the management and governance of individual lives.

The global picture represented by a policy of life aimed at a redefinition of the mechanisms and aims of medicine was specified and analysed in detail by Rose who chose to conduct a more thorough analysis of the changes in the relation between politics and life. Aware that politics has long been concerned with the vitality of the governed, Rose identified at least three phases in the history of this relation. A first phase, in which politics captured life as a set of statistically observable phenomena—birth, life, diseases, epidemics, mortality and so on—and generated a conception of public health aimed at the external conditions in which life develops (Rose, 2007, 3). A second phase, identified at the beginning of the twentieth century, in which these statistical and genetic concerns became linked to a conception of the hereditary nature and the biology of populations that fundamentally changed its meaning (ibid., 4). These concerns led to the birth of eugenics: power dealt with the internal conditions of life development and began to govern them. Finally, in the third and current phase, politics is not content with observing, ordering and classifying data concerning the life of the population, nor even reading them through the prism of a particular conception of heredity, but uses this knowledge to control, direct and shape individuals and their lives, starting therefore from their biological substance: politics, as Rose suggests, is from now on the politics of "life itself" (ibid., 3). In conclusion, politics is currently concerned with our growing ability to control, manage, design, shape and modulate the vital capacities of human beings as living creatures. In the twentieth century, Rose claims, there was a more thorough investigation of biopolitics: from now on the life we seek to govern is less and less inscribed on the surface of the body and is increasingly embedded in the molecular depths of its organic processes (ibid., 4). Following Rose's indications, it seems wrong to think that transhumanism and the manipulation of human biology in general can be designated as eugenic practices: their internal dynamic is aimed more at the creation of a new form of humanity, over which to gain control, than at ways of discriminating against sectors of the population.

This final and contemporary phase of the complex relation between politics and life is characterised by five features. First, a molecularisation of medicine, which increasingly deals with invisible processes at the subcorporeal level. Second, a process of optimisation, which is the side of the coin whose obverse characterises humanity as naturally exposed to disease and disability. Third, the emergence of new forms of subjectification that take place in relation to one's own soma. According to Rose, morality is increasingly becoming a field that establishes rights and duties, right and wrong, good and bad, in relation to an ideal of a good life characterised by a state of health. In contrast to what traditionally happens when a person's good life or morality is measured in relation to compliance with criteria of justice, we are witnessing the emergence of a somatic ethic that assesses behaviour against an ideal of good health defined in medical terms, whereby "good" or "just" individuals are those who adopt lifestyles marked by the pursuit and achievement of health (Rose, 2007, 6). Fourth, as a natural consequence of the above, an increased somatic competence is developing, leading to new forms of governance and, at the same time, the emergence of numerous sub-professions related to the field of health. Different categories of hyper-specialists of the body have appeared, whose role does not seem different from that of "shepherds", who act as an interface between an increasingly arcane medical-biological world and users increasingly disoriented by the innovative solutions being proposed to them. Finally, there is the development of a bioeconomy and economies of vitality, in which the surplus value is created by life and its vital processes. Of these five features, the ones that offer the most relevant interpretation are the optimisation of individual lives and the emergence of new forms of subjectification linked to the body, which thus becomes the area in which a new form of ethics can be implemented.

Optimisation has two dimensions: susceptibility and enhancement. These are mutually supportive and complementary features and it is interesting to see their interaction and effects. By susceptibility, Rose means a predisposition to disease, an absence of pathological symptoms that somehow conceals the potential onset of disease. The medical approach needed to tackle these possibilities, predisposed to intervene before the onset of a pathology, works in a totally different way from traditional medicine. By proposing to act before the outbreak of disease and even before its warning symptoms appear, its purpose will no longer be therapeutic, but defensive. The building up of these defences to prevent the onset of disease has a dual effect: it must optimise "the life chances of the individual" and, at the same time, it transforms all individuals into pre-patients that are constantly subjected to an indefinite extension of the powers of medical expertise (Rose, 2007, 19–20). Obviously, the pervasive procedures of this quest for health and the acceptance of its value give rise to views and practices that have a tendency to become totalitarian (ibid., 94). The inability to understand people who choose, for example, to have children who will acquire their own disabilities (Franssen, 2014; Sandel, 2004, 1; Savulescu, 2002) or, much more trivially, the many others who passively accept their pathology or completely ignore it, that is behaving as if they were healthy and refusing to undergo a subjectification that pathologises them, even refusing to explain or justify their behaviour, is significant of this obligation. Evidently, this refusal to rationalise their behaviour is interpreted as a symptom of irresponsibility: the individual is blamed, but it is precisely this attribution of blame that is the symptom of a medicalising system that acts in the context of this attempt to make individuals responsible for the maintenance or achievement of health (Rose, 2007, 94). In conclusion, with reference to Foucault's distinction between the determination of the ethical substance, the type of moral work required and the purpose of adaptation to moral imperatives, it could be said that this new morality makes "soma" the ethical substance on which the subject must work to prolong a healthy life as far as possible (Rose, 2007, 95).

The other side of the same coin is the trend towards what Rose calls optimisation, the extreme form of which is enhancement and represents the pivotal concept of transhumanism. Enhancement is a strategy whose effects go far beyond its explicit aims. In particular, this arguably pathological attention to the body has brought about an ethic centred on values such as physical fitness, health and a healthy lifestyle, which many view as signs of the person's moral value. Consequently, illness, pathologies, disorders of all kinds and unhealthy behaviour are signs of personal immorality (ibid., 25). Not only has our relationship with health changed but, due to biomedicine, so have the values attributed to this state and the judgement of individual behaviour. This somatisation of ethics has decreed a change in the way we perceive ourselves as individuals and has shifted all the weight of subjectification onto the individual's biological aspect which is as far removed as possible from the effects of chance. If the process of self-constitution is organised around an ideal of health, clearly everything that escapes its control generates anxiety and fear because it cannot be fully dominated. However, alongside anxiety and fear, there is a genesis of hope and expectation in a future in which morbidity and mortality will be eliminated (ibid., 27).

Rose maintains that this economics of hope can be defined overall as "ethopolitics" (ibid.). In other words, he states that we have entered an ethopolitical phase in which the governance and management of individual lives is performed in a different way from what was achieved in the simple biopolitical context described by Foucault. In this phase, politics uses the ethical principles through which one's own body is judged, managed and controlled, rather than the lives of the people themselves, as a tool of governance. Rose's analyses seem very significant and help us to understand why transhumanism could develop and why the criticisms directed at the dynamics of improvement that are proposed appear vain and very often wrong, counterproductive and contradictory. The aims of the transhumanist project, health, strength, youth and immortality, are the same as those that, according to Rose, are part of the imaginary and of the sciences of twentieth and twenty-first-century Western societies. This convergence shows that the positive values conveyed by transhumanism are also the values and imperatives through which individuals are governed (ibid.).

Agreeing with Harris that enhancement is a moral obligation or following Bostrom in his steadfast intent to become a posthumanist—do these not constitute individual choices as values and social imperatives? When Savulescu and Sandberg describe the economic benefits of empowerment and Buchanan even proposes to set up bodies and structures dedicated to the management and distribution of enhancement techniques, are they not taking a step from the optional to the imperative, are they not drawing a line separating a better and progressive society from one that has abandoned all willingness to advance and, in so doing, are they not indicating imperatives to be respected?

The use of biopolitical categories has allowed us to evaluate in a more adequate and in-depth way political and social consequences and modifications that, inherent in the contemporary world, are taken to a further level by transhumanism. The impact of transhumanism cannot be measured only at the level of individual dynamics that unduly reduce its importance, as if it were an individual improvement programme that pushes to produce the best version of oneself. Its effects lie at different and incomparable levels that are not often described in all their breadth and depth.

However, it is not only in the passage from the individual to the social that transhumanism reveals its aims, but it is also in the implicit suggestion that people who work for their own enhancement are morally better that the function of transhumanism is manifested. It is by stirring up hidden tendencies, but no less strong because of it, by instigating ancestral fears, but no less modern because of it, by pursuing the most violent desires, but no less shared because of it, that transhumanism makes us believe that enhancement is exactly what we seek, indeed it is what we must seek in order to be better. Nourishing our narcissism, it pushes us towards the pool of water in which we study ourselves more and more closely, and then it gives us a little push making us fall into that pond from which we will no longer be able to get out. Concentrating on what we thought was our reflection, we failed to realise that the water, which at first seemed so clear and attractive, was in fact murky and noisome. And our reflection may no longer be that of a "*Citius, Fortius, Altius*" being, but of an unspeakable monster.

References

Arendt, H. (1998). *The human condition* (2nd ed.). University of Chicago Press.

Buchanan, A. (2011b). *Beyond humanity?* Oxford University Press.

Foucault, M. (1994a). Crisis de la médecine ou crise de l'antimédecine? In M. Foucault (Ed.), *Dits et écrits* (Vol. III, pp. 40–57). Gallimard.

Foucault, M. (1994b). L'incorporation de l'hôpital dans la technologie moderne. In M. Foucault (Ed.), *Dits et écrits* (Vol. III, pp. 508–521). Gallimard.

Foucault, M. (1998). *The history of sexuality: Vol 1, The will to knowledge.* Penguin.

Foucault, M. (2001a). *Essential works of Foucault 1954–1984: Vol 3, Power.* The New Press.

Foucault, M. (2001b). The politics of health in the eighteenth century. In *Essential works of Foucault 1954–1984: Vol 3, Power* (pp. 90–105). The New Press.

Foucault, M. (2001c). The birth of social medicine. In *Essential works of Foucault 1954–1984: Vol 3, Power* (pp. 134–156). The New Press.

Foucault, M. (2007). *Security, territory, population: Lectures at the College De France, 1977–78.* Palgrave Macmillan.

Franssen, T. (2014). Prometheus descends: Disabled or enhanced? John Harris, human enhancement, and the creation of a new norm. In M. Eilers, K. Grüber, & C. Rehmann-Sutter (Eds.), *The human enhancement debate and disability: New bodies for a better life* (pp. 161–182). Palgrave Macmillan.

Heidegger, M. (1991). *Nietzsche* (Vols. III & IV). Harper & Collins.

Hughes, J. (2004). *Citizen cyborg: Why democratic societies must respond to the redesigned human of the future.* Westview Press Books.

More, M., & Vita-More, N. (2013). *The transhumanist reader.* Wiley Blackwell.

Reiner, P. (2013). The biopolitics of cognitive enhancement. In E. Hildt & A. Franke (Eds.), *Cognitive enhancement: An interdisciplinary perspective* (pp. 189–200). Springer.

Rose, N. (2007). *The politics of life itself: Biomedicine, Power, and subjectivity in the twenty-first century.* Princeton University Press.

Sandberg, A., & Savulescu, J. (2011). The social and economic impacts of cognitive enhancement. In J. Savulescu, R. ter Meulen, & G. Kahane (Eds.), *Enhancing human capacity* (pp. 92–112). Wiley-Blackwell.

Sandel, M. (2004). The case against perfection. *The Atlantic Monthly*, 1–10.

Savulescu, J. (2002). Deaf Lesbians, 'Designer Disability', and the future of medicine. *BMJ, 325*(7367), 771–773.
Savulescu, J., & Bostrom, N. (2009). *Human enhancement*. Oxford University Press.
Staman, J., Dijstelbloem, H., & Smits, M. (2008). Afterword. In L. Zonneveld (Ed.), *Reshaping the human condition: Exploring human enhancement* (pp. 149–157). Rathenau Institute.

Bibliography

Adams, D. (1979). *The Hitchhiker's Guide to the Galaxy*. Pan Books.

Adorno, F. P., & Goffi, J.-Y. (2015). Le Transhumanisme. *Ethique, Politique, Religions, 6*(1).

Agar, N. (2004). *Liberal Eugenics*. Blackwell.

Agar, N. (2010). *Humanity's End*. MIT Press.

Agar, N. (2014). *Truly human enhancement: A philosophical defense of limits*. MIT Press.

Agar, N. (2015). Moral bioenhancement is dangerous. *Journal of Medical Ethics, 41*(4), 343–345.

Alexander, B. (2003). *Rapture: A raucous tour of cloning, transhumanism, and the new era of immortality*. Basic Books.

Alexandre, L. (2011). *La mort de la mort*. Lattès.

Allenby, B., & Daniel, S. (2011). *The techno-human condition*. MIT Press.

Appleyard, B. (2007). *How to live forever or die trying*. Simon & Schuster.

Arendt, H. (1998). *The human condition* (2nd ed.). University of Chicago Press.

Aristotle. (n.d.). *Nicomachean ethics* (W. D. Ross, Trans.). The Internet Classic Archive. http://classics.mit.edu//Aristotle/nicomachaen.html

F. P. Adorno, *The Transhumanist Movement*,
https://doi.org/10.1007/978-3-030-82423-5

Arrhenius, G. (2011). Life extension versus replacement. In J. Savulescu, R. ter Meulen, & G. Kahane (Eds.), *Enhancing human capacity* (pp. 368–385). Wiley-Blackwell.

Arthur, W. B. (2009). *The nature of technology: What it is and how it evolves.* Free Press.

Bacon, F. (1924). *New Atlantis.* Clarendon Press.

Bacon, F. (2006). De viis mortis, et de senectute retardanda, atque instaurandis viribus. In G. Rees & L. Jardin (Eds.), *The Oxford Francis Bacon* (Vol. IV, pp. 270–359). Clarendon Press.

Badmington, N. (2000). *Posthumanism.* Palgrave.

Baertschi, B. (2009). Devenir un être humain accompli: idéal ou cauchemar. In J.-N. Missa & L. Perbal (Eds.), *«Enhancement». Ethique et philosophie de la médecine d'amélioration* (pp. 79–95). Vrin.

Baertschi, B. (2014). Neuromodulation in the service of moral enhancement. *Brain Topography, 27*(1), 63–71.

Bailey, R. (2005). *Liberation biology: The scientific and moral case for the biotech revolution.* Prometheus Books.

Baily, H. W., & T. K. Casey. (2005). *Is human nature obsolete? Genetics, bioengineering and the future of the human condition.* MIT Press.

Barazzetti, G. (2011). Looking for the fountain of youth: Scientific, ethical, and social issues in the extension of human lifespan. In J. Savulescu, R. ter Meulen, & G. Kahane (Eds.), *Enhancing human capacity* (pp. 335–349). Wiley-Blackwell.

Barazzetti, G., & Reichlin, M. (2011). Life extension and personal identity. In J. Savulescu, R. ter Meulen, & G. Kahane (Eds.), *Enhancing human capacity* (pp. 398–409). Wiley-Blackwell.

Bashford, A. (2013). Julian Huxley's transhumanism. In M. Turda (Ed.), *Crafting humans: From genesis to eugenics and beyond* (pp. 53–167). National Taiwan University Press.

Bateman, S., & Gayon, J. (2012). L'amélioration humaine: trois usages, trois enjeux. *Médecine/sciences, 10*(28), 887–891.

Bateman, S., Gayon, J., Allouche, S., Goffette, J., & Marzano, M. (2015). *Inquiring into human enhancement: Interdisciplinary and international perspectives.* Palgrave Macmillan.

Bauman, Z. (1992). *Mortality, immortality and other life strategies.* Polity Press.

Baumgartner, T., Heinrichs, M., Vonlanthen, A., Fischbacher, U., & Fehr, E. (2008). Oxytocin shapes the neural circuitry of trust and trust adaptation in humans. *Neuron, 58*(49), 639–650.

Beauchamp, T. L. (2015). Are we unfit for the future? *Journal of Medical Ethics, 41*(4), 346–348.

Becker, E. (1973). *The denial of death*. Simon & Schuster.

Berghmans, R., ter Meulen, R., Malizia, A., & Vos, R. (2011). Scientific, ethical, and social issues in mood enhancement. In J. Savulescu, R. ter Meulen, & G. Kahane (Eds.), *Enhancing human capacity* (pp. 153–165). Wiley-Blackwell.

Berlin, I. (2002). *Liberty*. Oxford University Press.

Bernal, J. D. (1969). *The world, the flesh and the devil: An enquiry into the future of the three enemies of the rational soul* (1929). Indiana University Press.

Bond, J. (2011). Enhancing human aging: The cultural and psychosocial context of lifespan extension. In J. Savulescu, R. ter Meulen, & G. Kahane (Eds.), *Enhancing human capacity* (pp. 435–452). Wiley-Blackwell.

Boorse, C. (1975). On the distinction between disease and illness. *Philosophy & Public Affairs, 51*, 49–68.

Boorse, C. (1977). Health as a theoretical concept. *Philosophy of Science, 44*(4), 542–573.

Borges, J.-L. (1957). *The Immortal*. In Borges J.-L. 1962. *Labyrinths and other stories*. New Directions.

Bostrom, N. (2002). Existential risks. *Journal of Evolution and Technology, 9*(1), 1–31.

Bostrom, N. (2005a). A history of transhumanist thought. *Journal of Evolution and Technology, 14*(1), 1–25.

Bostrom, N. (2005b). In defense of posthuman dignity. *Bioethics, 19*, 202–214.

Bostrom, N. (2005c). Transhumanist values. *Review of Contemporary Philosophy, 4*, 87–101.

Bostrom, N. (2005d). The fable of the dragon tyrant. *Journal of Medical Ethics, 31*, 273–277.

Bostrom, N. (2008a). Letter from Utopia. *Journal of Evolution and Technology, 19*(1), 67–72.

Bostrom, N. (2008b). Smart policy: Cognitive enhancement in the public interest. In L. Zonneveld (Ed.), *Reshaping the human condition: Exploring human enhancement*. Rathenau Institute.

Bostrom, N. (2009a). Why I want to be a posthuman when I grow up. In A. M. Cutter, B. Gordijn, G. E. Marchant, A. Pompidou, & R. Chadwick (Eds.), *Medical enhancement and posthumanity* (pp. 107–136). Springer.

Bostrom, N. (2009b). Dignity and enhancement. In B. Lanigan (Ed.), *Human dignity and bioethics* (pp. 123–144). Nova Science Pub Inc.

Bostrom, N. (2014). *Superintelligence*. Oxford University Press.

Bostrom, N., & Ord, T. (2006). The reversal test: Eliminating status quo bias in applied ethics. *Ethics, 116*(4), 656–679.

Bostrom, N., & Roache, R. (2007). Ethical issues in human enhancement. In J. Ryberg, T. S. Petersen, & C. Wolf (Eds.), *New waves in applied ethics* (pp. 120–152). Palgrave Macmillan.

Bostrom, N., & Roache, R. (2011). Smart policy: Cognitive enhancement and the public interest. In J. Savulescu, R. ter Muelen, & G. Kahane (Eds.), *Enhancing human capabilities* (pp. 138–149). Wiley-Blackwell.

Bostrom, N., & Sandberg, A. (2009a). Cognitive enhancement: Methods, ethics, regulatory challenges. *Science and Engineering Ethics, 15*(3), 311–341.

Bostrom, N., & Sandberg, A. (2009b). The wisdom of nature: An evolutionary heuristic for human enhancement. In J. Savalescu & N. Bostrom (Eds.), *Human enhancement* (pp. 375–416). Oxford University Press.

Bostrom, N., & Savulescu, J. (2009). Human enhancement ethics: The state of the debate. In J. Savulescu & N. Bostrom (Eds.), *Human enhancement* (pp. 1–22). Oxford University Press.

Bova, B. (1998). *Immortality: How science is extending your life span and changing the world*. Avon Books.

Braidotti, R. (2013). *The posthuman.* Polity Press.

Brey, P. (2009). Human enhancement and personal identity. In J.-K.B. Olsen, E. Selinger, & S. Riis (Eds.), *New Waves in Philosophy of Technology* (pp. 169–185). Palgrave Macmillan.

Brey, P. (2014). From moral agents to moral factors: The structural ethics approach. In P. Kroes & P.-P. Verbeek (Eds.), *The moral status of technical artefacts* (pp. 125–147). Springer.

Brock, D. W. (2009). Is selection of children wrong? In J. Savulescu & N. Bostrom (Eds.), *Human enhancement* (pp. 251–276). Oxford University Press.

Brulde, B. (2011). Is mood enhancement a legitimate goal of medicine? In J. Savulescu, R. ter Meulen, & G. Kahane (Eds.), *Enhancing human capacity* (pp. 218–229). Wiley-Blackwell.

Brunon-Ernst, A. (2012). *Utilitarian biopolitics: Bentham, Foucault and modern power*. Pickering & Chatto Ltd.

Bublitz, J. C., & Merkel, R. (2009). Autonomy and authenticity of enhanced personality traits. *Bioethics, 23*(6), 360–374.

Buchanan, A. (2011a). *Better than human: The promise and perils of enhancing ourselves*. Oxford University Press.

Buchanan, A. (2011b). *Beyond humanity?* Oxford University Press.

Buchanan, A., Brock, D. W., Daniels, N., & Wikler, D. (2001). *From chance to choice: Genetics and justice*. Cambridge University Press.

Cabrera, L. Y. (2015). *Rethinking human enhancement: Social enhancement and emergent technologies*. Palgrave Macmillan.

Caldera, E. O. (2008). Cognitive enhancement and theories of justice: Contemplating the malleability of nature and self. *Journal of Evolution and Technology, 18*(1), 116–123.

Campbell, E. J. M., Scadding, J. G., & Roberts, R. S. (1979). The concept of disease. *The British Medical Journal, 2*(6193), 757–762.

Campbell, N., O'Driscoll, A., & Saren, M. (2010). The posthuman: The end and the beginning of the human. *Journal of Consumer Behaviour, 9*, 86–101.

Camus, A. (1954). *The Rebel: An essay on man in revolt*. A.A. Knopf.

Canguilhem, G. (1989). Machine et organisme. In *La connaissance de la vie* (pp. 101–127). Vrin.

Canguilhem, G. (1991). *The normal and the pathological*. Zone Books.

Canton, J. (2004). Designing the future: NBIC Technologies and human performance enhancement. *Annals of the New York Academy of Sciences, 1013*, 186–198.

Caplan, A. (2009). Good, better, or best? In J. Savulescu & N. Bostrom (Eds.), *Human enhancement* (pp. 199–210). Oxford University Press.

Carnevale, A., & Battaglia, F. (2014). A reflexive approach to human enhancement: Some philosophical considerations. In F. Lucivero & A. Vedder (Eds.), *Beyond therapy v. enhancement? Multidisciplinary analysis of a heated debate* (pp. 95–116). Pisa University Press.

Carrel, A. (1935). *L'homme, cet inconnu*. Plon.

Carter, J. A., & Gordon, E. C. (2015). On cognitive and moral enhancement: A reply to Savulescu and Persson. *Bioethics, 3*(29), 153–161.

Cartwright, L., & Goldfarb, B. (2006). On the subject of neural and sensory prostheses. In M. E. Smith & J. Morra (Eds.), *The prosthetic impulse: From a posthuman present to a biocultural future* (pp. 125–154). MIT Press.

Casal, P. (2015). On not taking men as they are: Reflections on moral bioenhancement. *Journal of Medical Ethics, 41*(4), 340–342.

Cassirer, E. (1948). *The renaissance philosophy of man*. Chicago University Press.

Cave, S. (2012). *Immortality: The quest to live forever and how it drives civilization*. Crown Publishers.

Chadwick, R. (2009). Therapy, enhancement and improvement. In B. Gordijn & R. Chadwick (Eds.), *Medical enhancement and posthumanity* (pp. 25–38). Springer.

Chan, S., & Harris, J. (2008). In support of Human Enhancement. *Studies in Ethics, Law, and Technology, 1*(1), 1–3.

Chan, S., & Harris, J. (2011). Moral enhancement and pro-social behaviour. *Journal of Medical Ethics, 37*(3), 130–131.

Charis Hall, M. (2017). *The bioethics of enhancement: Transhumanism, disability, and biopolitics*. Lexington Books.

Chatterjee, A. (2004). Cosmetic neurology the controversy over enhancing movement, mentation, and mood. *Neurology, 63*(6), 968–974.

Christman, J. (1989). *The Inner Citadel: Essays on individual autonomy*. Oxford University Press.

Chu, T. (2014). *Human purpose and transhuman potential*. Origin Press.

Churchland, P. (2012). *Braintrust: What neuroscience tells us about morality*. Princeton University Press.

Cioran, E. (1968). *The temptation to exist*. Arcade.

Clark, A. (1998). *Being there: Putting brain, body, and world together again*. MIT Press.

Clark, A. (2004). *Natural-born cyborgs*. Oxford University Press.

Clark, A. (2008). *Supersizing the mind: Embodiment, action, and cognitive extension*. Oxford University Press.

Clark, A. (2010). Memento's revenge: The extended mind, extended. In R. Menary (Ed.), *The extended mind* (pp. 43–66). MIT Press.

Clark, A., & Chalmers, D. (1998). The extended mind. *Analysis, 581*, 7–19.

Clarke, S. (2011). Cognitive bias and collective enhancement. In J. Savulescu, R. ter Meulen, & G. Kahane (Eds.), *Enhancing human capacity* (pp. 127–137). Wiley-Blackwell.

Clarke, S., Savulescu, J., Coady, T., Giubilini, A., & Sanyal, S. (2016). *The ethics of human enhancement: Understanding the debate*. Oxford University Press.

Clynes, M. E., & Kline, N. (1960). Cyborgs and space. *Astronautics, 13*, 27–31.

Coeckelbergh, M. (2013a). *Human Being @ Risk: Enhancement, technology, and the evaluation of vulnerability transformations*. Springer Netherlands.

Coeckelbergh, M. (2013b). Enhancement and the vulnerable body: Questioning some philosophical assumptions. In F. Lucivero & A. Vedder (Eds.), *Beyond therapy v. enhancement* (pp. 15–26). Pisa University Press.

Coenen, C. (2014). Transhumanism and its genesis: The shaping of human enhancement discourse by visions of the future. *Humana Mente: Journal of Philosophical Studies, 26*, 35–58.

Coenen, C., Schuijff, M., & Smits, M. (2011). The politics of human enhancement and the European Union. In J. Savulescu, R. ter Meulen, & G. Kahane (Eds.), *Enhancing human capacity* (pp. 521–534). Wiley-Blackwell.

Coenen, C., Schuijff, M., Smits, M., Klaassen, P., Hennen, L., Rader, M., & Wolbring, G. (2009). *Human enhancement study*. European Parliament. Science and Technology Assessment Group.

Coeurnelle, D. (2013). *Et si on arrêtait de vieillir!* Fyp éditions.

Coeurnelle, D., & Roux, M. (2016). *Technoprog. Le transhumanisme au service du progrès social*. Fyp éditions.

Comte, A. (1996). *Philosophie des sciences*. Gallimard.

Conrad, P. (2007). *The medicalization of society: On the transformation of human conditions into treatable disorders*. The Johns Hopkins University Press.

Cooper, M. (2008). *Life as surplus: Biotechnology and capitalism in the neoliberal era*. University of Washington Press.

Cordeiro, J. L. (2019). The boundaries of the human: From humanism to transhumanism. In N. Lee (Ed.), *The transhumanism handbook* (pp. 63–74). Springer.

Critical Art Ensemble. (1998). *Flesh Machine: Cyborgs, Designer Babies, and New Eugenic Consciousness*. Autonomedia.

Crockett, M. J. (2014). Moral bioenhancement: A neuroscientific perspective. *Journal of Medical Ethics, 40*(6), 370–371.

Damasio, A. (1994). *Descartes' error: Emotion, reason, and the human brain*. Putnam Publishing.

Damour, F., & Doat, D. (2018). *Transhumanisme. Quel avenir pour l'humanité?* Le Cavalier Bleu.

Daniels, N. (1985). *Just health care*. Cambridge University Press.

Daniels, N. (1992). Growth hormone therapy for short stature: Can we support the treatment/enhancement distinction? *Growth: Genetics & Hormones, 8*, 46–48.

Daniels, N. (2000). Normal functioning and the treatment-enhancement distinction. *Cambridge Quarterly of Healthcare Ethics, 9*(3), 309–322.

Daniels, N. (2008). *Just health*. Cambridge University Press.

Daniels, N. (2009). Can anyone really be talking about ethically modifying human nature? In J. Savulescu & N. Bostrom (Eds.), *Human enhancement* (pp. 25–42). Oxford University Press.

Dard, O., & Moatti, A. (2016). Aux origines du mot 'transhumanisme.' *Futuribles, 413*, 154–165.

Davies, T. (1997). *Humanism*. Routledge.

Dawkins, R. (1986). *The Blind Watchmaker: Why the evidence of evolution reveals a universe without design.* Norton & Co.

de Condorcet, M. (1955). *Sketch for a historical picture of the progress of the human mind*. Weidenfeld & Nicolson.

De Grey, A. (2004a). Escape velocity: Why the prospect of extreme human life extension matters now. *PLOS Biology, 2*(6), 0723–0726.

De Grey, A. (2004b). The war on aging. In M. R. Rose, et al. (Eds.), *The scientific conquest of death: Essays on infinite lifespans* (pp. 29–46). Libros en Red.

De Grey, A., & Rae, M. (2007). *Ending aging*. Saint Martin's Press.

DeGrazia, D. (2000). Prozac, enhancement, and self-creation. *The Hastings Center Report, 30*(2), 34–40.

DeGrazia, D. (2005). Enhancement technologies and human identity. *Journal of Medicine and Philosophy, 30*(3), 261–283.

DeGrazia, D. (2014). Moral enhancement, freedom, and what we (should) value in moral behaviour. *Journal of Medical Ethics, 40*(6), 361–368.

Deitch, J. (1992). *Post human: Exhibit catalog essay 1992.* FAE Musée d'Art Contemporain Pully.

Delfin, D. (2019). An artist's creative process: A model for conscious evolution. In N. Lee (Ed.), *The transhumanism handbook* (pp. 587–602). Springer.

Del Val, J., & Sorgner, S. (2010). *Metahumanist Manifesto.* www.metahumanism.eu

Deprez, S. (2019). Le transhumanisme peut-il échapper au libertarianisme? Le cas de James Hugues. *Raisons Politiques, 74*(2), 29–50.

Devereaux, M. (2008). Cosmetic surgery. In B. Gordijn & R. Chadwick (Eds.), *Medical enhancement and posthumanity* (pp. 169–174). Springer.

Dickel, S., & Frewer, A. (2014). Life extension: Eternal debates on immortality. In R. Ranisch & S. L. Sorgner (Eds.), *Post- and transhumanism: An introduction* (pp. 119–131). Peter Lang.

Douglas, T. (2008). Moral enhancement. *Journal of Applied Philosophy, 25*(3), 228–245.

Douglas, T. (2011a). Moral enhancement. In J. Savulescu, R. ter Meulen, & G. Kahane (Eds.), *Enhancing human capacity* (pp. 467–485). Wiley-Blackwell.

Douglas, T. (2011b). Human enhancement and supra-personal moral status. *Philosophical Studies, 162*(3), 473–497.

Douglas, T. (2013). Moral enhancement via direct emotion modulation: A reply to John Harris. *Bioethics, 27*(3), 160–168.

Doyle, J. (2018). *What does it mean to be human? Life, death, personhood and the transhumanist movement*. Springer.

Drexler, E. (1986). *Engines of creation.* Anchor Books.

Drexler, E. (1992). *Nanosystems: Molecular machinery, manufacturing, and computation.* Wiley.

Dupuy, J.-P. (2011). Cybernetics is antihumanism: Advanced technologies and the rebellion against the human condition. In G. R. Hansell & W. J. Grassie (Eds.), *H+ transhumanism and its critics* (pp. 101–110). Metanexus Institute.

Dworkin, R. (2002). *Sovereign virtue: The theory and practice of equality.* Harvard University Press.

Earp, B. D., Sandberg, A., Kahane, G., & Savulescu, J. (2014). When is diminishment a form of enhancement? Rethinking the enhancement debate in biomedical ethics. *Frontiers in Systems Neuroscience, 81*(2), 1–8.

Edgar, A. (2009). The hermeneutic challenge of genetic engineering: Habermas and the transhumanists. *Medicine, Health Care and Philosophy, 12*(2), 157–167.

Eilers, M., Grüber, K., & Rehmann-Sutter, C. (2014). *The human enhancement debate and disability: New bodies for a better life.* Palgrave Macmillan.

Elliott, C. (1999). *A philosophical disease.* Routledge.

Elliott, C. (2004). *Better than well: American medicine meets the American dream.* Norton & Company.

Engelhardt, H. T. (1976). Is there a philosophy of medicine? *Proceedings of the Biennial Meeting of the Philosophy of Science Association, 2*, 94–108.

Engelhardt, H. T. (1991). *Bioethics and secular humanism: The search for a common morality.* SCM Press-Trinity Press International.

Esfandiary FM (FM-2030). (1977a). *Up-wingers: A Futurist Manifesto.* Popular Library.

Esfandiary FM (FM-2030). (1977b). *Telespheres.* Popular Library.

Esfandiary FM (FM-2030). (1978). *Optimism one* (1970). Popular Library.

Esfandiary FM (FM-2030). (1989). *Are you a transhuman? Monitoring and stimulating your personal rate of growth in a rapidly changing world.* Warner Books.

Ettinger, R. (1962). *The prospect of immortality.* Ria University Press.

Ettinger, R. (1972). *Man into superman.* St. Martin's Press.

Farah, M. J. (2004). Emerging ethical issues in neuroscience. *Nature Neuroscience, 5*(11), 1123–1130.

Farah, M. J. (2010). *Neuroethics: An introduction with readings.* MIT Press.

Farah, M. J., & Heberlein, A. S. (2007). Personhood and neuroscience: Naturalizing or nihilating? *The American Journal of Bioethics, 71*, 37–48.

Farah, M. J., Illes, J., Cook-Deegan, R., Gardner, H., Kandel, E., King, P., Parens, E., Sahakian, B., & Wolpe, P. W. (2004). Neurocognitive enhancement: What can we do and what should we do? *Nature Reviews Neuroscience, 5*(5), 421–425.

Farah, M. J., & Wolpe, P. R. (2004). Monitoring and manipulating brain function: New neuroscience technologies and their ethical implications. *Hastings Center Report, 34*(3), 35–45.

Fenton, E. (2010). The perils of failing to enhance: A response to Persson and Savulescu. *Journal of Medical Ethics, 36*(3), 148–151.

Fernández-Armesto, F. (2005). *So you think you're human? A brief history of humankind*. Oxford University Press.

Ferrando, F. (2019). *Philosophical posthumanism*. Bloomsbury Academic.

Flanigan, J., & Price, T. L. (2018). *The ethics of ability and enhancement*. Palgrave Macmillan.

Foddy, B. (2011). Enhancing skill. In J. Savulescu, R. ter Meulen, & G. Kahane (Eds.), *Enhancing human capacity* (pp. 313–325). Wiley-Blackwell.

Foucault, M. (1970). *The order of things: An archaeology of the human sciences*. Pantheon Books.

Foucault, M. (1994a). Crisis de la médecine ou crise de l'antimédecine? In M. Foucault (Ed.), *Dits et écrits* (Vol. III, pp. 40–57). Gallimard.

Foucault, M. (1994b). L'incorporation de l'hôpital dans la technologie moderne. In M. Foucault (Ed.), *Dits et écrits* (Vol. III, pp. 508–521). Gallimard.

Foucault, M. (1994c). La politique de la santé au XVIIIème siècle. In M. Foucault (Ed.), *Dits et écrits* (Vol. III, pp. 725–741). Gallimard.

Foucault, M. (1997). *Essential works of Foucault, 1954–1984: Vol 1, Ethics*. The New Press.

Foucault, M. (1998a). *The history of sexuality: Vol 1, The will to knowledge*. Penguin.

Foucault, M. (1998b). *Essential works of Foucault, 1954–1984: Vol 2, Aesthetics, method, and epistemology*. The New Press.

Foucault, M. (2001a). *Essential works of Foucault 1954–1984: Vol 3, Power*. The New Press.

Foucault, M. (2001b). The politics of health in the eighteenth century. In *Essential works of Foucault 1954–1984: Vol 3, Power* (pp. 90–105). The New Press.

Foucault, M. (2001c). The birth of social medicine. In *Essential works of Foucault 1954–1984: Vol 3, Power* (pp. 134–156). The New Press.

Foucault, M. (2007). *Security, territory, population: Lectures at the College De France, 1977–78*. Palgrave Macmillan.

Frankford, D. M. (1998). The treatment/enhancement distinction as an armament in the policy wars. In E. Parens (Ed.), *Enhancing human traits: Ethical and social implications* (pp. 70–94). Georgetown University Press.

Franssen, T. (2014). Prometheus descends: Disabled or enhanced? John Harris, human enhancement, and the creation of a new norm. In M. Eilers, K. Grüber, & C. Rehmann-Sutter (Eds.), *The human enhancement debate and disability: New bodies for a better life* (pp. 161–182). Palgrave Macmillan.

Freedman, C. (1998). Aspirin for the mind? Some ethical worries about psychopharmacology. In E. Parens (Ed.), *Enhancing human traits: Ethical and social implications* (pp. 135–150). Georgetown University Press.

Freitas, R. A. (2007). Personal choice in the coming era of nanomedicine. In F. Allhoff, P. Lin, J. Moor, & J. Weckert (Eds.), *Nanoethics: The ethical and social implications of nanotechnology* (pp. 161–172). Wiley.

Fröding, B. (2013). *Virtue ethics and human enhancement*. Springer.

Fukuyama, F. (2002). *Our posthuman future: Consequences of the biotechnology revolution.* Farrar, Straus and Giroux.

Fukuyama, F. (2004). Transhumanism: The world's most dangerous idea. *Foreign Policy, 144*, 42–43.

Fuller, S., & Lipinska, V. (2014). *The proactionary imperative: A foundation for transhumanism.* Palgrave Macmillan.

Garreau, J. (2005). *Radical evolution: The promise and peril of enhancing our minds, our bodies and what means to be human.* Doubleday.

Gazzaniga, M. S. (2005). *The ethical brain.* Dana Press.

Gems, D. (2003). Is more life always better? The new biology of aging and the meaning of life. *Hastings Center Report, 4*(33), 31–39.

Giubilini, A., & Sanyal, S. (2016). Challenging human enhancement. In S. Clarke, J. Savulescu, T. Coady, A. Giubilini, & S. Sanyal (Eds.), *The ethics of human enhancement: Understanding the debate* (pp. 1–24). Oxford University Press.

Glannon, W. (2002). Identity, prudential concern, and extended lives. *Bioethics, 3*(16), 266–283.

Glannon, W. (2006). Neuroethics. *Bioethics, 1*(20), 37–52.

Glannon, W. (2008a). *Bioethics and the brain.* Oxford University Press.

Glannon, W. (2008b). Decelerating and arresting human aging. In B. Gordijn & R. Chadwick (Eds.), *Medical enhancement and posthumanity* (pp. 175–190). Springer.

Glannon, W. (2013). *Brain, body and mind: Neuroethics with a human face*. Oxford University Press.
Glenn, L. M., & Dvorsky, G. (2010). Dignity and agential realism: Human, posthuman, and nonhuman. *The American Journal of Bioethics, 10*(7), 57–58.
Glover, J. (1984). *What sort of people should there be? Genetic engineering, brain control, and their impact on our future world*. Penguin.
Glover, J. (1988). *I: The philosophy and psychology of personal identity*. Penguin.
Goffette, J. (2006). *Naissance de l'anthropotechnie. De la médecine au modelage de l'humain*. Vrin.
Goffette, J. (2009). Modifier les humains: Anthropotechnie versus médecine. In J.-N. Missa & L.Perbal (Eds.), *«Enhancement». Ethique et philosophie de la médecine d'amélioration* (pp. 49–63). Vrin.
Goffi, J.-Y. (2009). Thérapie, augmentation et finalité de la médecine. In J. N. Missa & L. Perbal (Eds.), *«Enhancement». Ethique et philosophie de la médecine d'amélioration* (pp. 97–120). Vrin.
Goffi, J.-Y. (2015). Aux origines contemporaines du transhumanisme. In F. P. Adorno & J.-Y. Goffi (Eds.), Le Transhumanisme. *Ethique, politique, religions, 1*(6), 17–32.
Goldman, A. I. (1993). Ethics and cognitive science. *Ethics, 1032*, 337–360.
Goodman, R. (2010). Cognitive enhancement, cheating, and accomplishment. *Kennedy Institute of Ethics Journal, 202*, 145–160.
Gordijn, B. (2005). Nanoethics: From Utopian dreams and apocalyptic nightmares towards a more balanced view. *Science and Engineering Ethics, 11*(4), 521–533.
Gordijn, B. (2006). Converging NBIC technologies for improving human performance: A critical assessment of the novelty and the prospects of the project. *The Journal of Law, Medicine & Ethics, 34*(4), 726–732.
Gordijn, B., & Chadwick, R. (2008). *Medical enhancement and posthumanity*. Springer.
Graham, E. L. (2002). *Representations of the post/human*. Rutgers University Press.
Graham, J., Haidt, J., & Nosek, B. A. (2009). Liberals and conservatives rely on different sets of moral foundations. *Journal of Personality and Social Psychology, 96*(5), 1029–1046.
Gray, C. H. (1995). *The cyborg handbook*. Routledge.
Gray, C. H. (2000). *Cyborg citizen: Politics in the posthuman age*. Routledge.

Greely, H., Sahakian, B., Harris, J., Kessler, R. C., Gazzaniga, M., Campbell, P., & Farah, M. J. (2008). Towards responsible use of cognitive-enhancing drugs by the healthy. *Nature, 456*(7223), 702–705.

Greely, H. T. (2010). Enhancing brains: What are we afraid of? *Cerebrum, 14*, 1–10.

Greely, H. T. (2011). Of nails and hammers: Human biological enhancement and U.S. Policy tools. In J. Savulescu, R. ter Meulen, & G. Kahane (Eds.), *Enhancing human capacity* (pp. 503–520). Wiley-Blackwell.

Habermas, J. (2003). *The future of human nature*. Polity.

Haidt, J. (2001). The emotional dog and its rational tail: A social intuitionist approach to moral judgment. *Psychological Review, 108*(4), 814–834.

Haisma, H. J. (2011). Physical enhancement. In J. Savulescu, R. ter Meulen, & G. Kahane (Eds.), *Enhancing human capacity* (pp. 259–265). Wiley-Blackwell.

Halberstam, J. M., & Livingston, I. (1995). *posthuman bodies*. Indiana University Press.

Haldane, J. B. S. (1923). *Daedalus or science and the future: A paper read to the heretics cambridge*. Dutton & Company.

Hansell, G. R., & Grassie, W. J. (2011). *H+ transhumanism and its critics*. Metanexus Institute.

Haraway, D. J. (1991). A cyborg manifesto: Science, technology, and socialist feminism in the late twentieth century. In *Simians, cyborgs and women: The reinvention of human nature*. Routledge.

Harnacke, C. (2013). Discussing enhancement and disability together. In F. Lucivero & A. Vedder (Eds.), *Beyond therapy enhancement* (pp. 69–94). Pisa University Press.

Harrington, A. (1969). *The immortalist*. Random House.

Harris, J. (1992). *Wonderwoman and superman: The ethics of human biotechnology*. Oxford University Press.

Harris, J. (2007). *Enhancing evolution*. Princeton University Press.

Harris, J. (2009). Enhancements are a moral obligation. In J. Savulescu & N. Bostrom, *Human enhancement* (pp. 131–154). Oxford University Press.

Harris, J. (2010). Taking the 'human' out of human rights. *Cambridge Quarterly of Healthcare Ethics, 20*(1), 9–20.

Harris, J. (2011). Moral enhancement and freedom. *Bioethics, 25*(2), 102–111.

Harris, J. (2013). Ethics is for bad guys! Putting the 'moral' into moral enhancement. *Bioethics, 27*(3), 169–173.

Harrison, P., & Wolyniak, J. (2015). The history of 'transhumanism.' *Notes and Queries, 62*(3), 465–467.

Hassan, I. (1977). Prometheus as performer: Toward a posthumanist culture? In M. Benamou & C. Caramello (Eds.), *Performance in postmodern culture* (pp. 201–217). Coda Press.

Hayles, N. K. (1999). *How we became posthuman: Virtual bodies in cybernetics, literature, and informatics*. University of Chicago Press.

Hayflick, L. (1994). *How and why we age*. Ballantine Books.

Hauskeller, M. (2013). *Better humans: Understanding the enhancement project*. Acumen.

Hauskeller, M. (2016). *Mythologies of transhumanism*. Palgrave Macmillan.

Hauskeller, M. (2015). Being good enough to prevent the worst. *Journal of Medical Ethics, 41*(4), 289–290.

Hempel, C. G. (1965). *Aspects of scientific explanation and other essays in the philosophy of science*. Free Press.

Herath, D., Kroos, C., & Stelarc. (2016). *Robots and art: Exploring an unlikely symbiosis*. Springer.

Herbrechter, S. (2013). *Posthumanism: A critical analysis*. Bloomsbury.

Hess, D. (1995). On low tech cyborg. In C. H. Gray (Ed.), *The cyborg handbook* (pp. 371–378). Routledge.

Heidegger, M. (1991). *Nietzsche* (Vols. III & IV). Harper & Collins.

Hildt, E., & Franke, A. (2013). *Cognitive enhancement: An interdisciplinary perspective*. Springer.

Holm, S. (2013). Does nanotechnology require a new 'nanoethics?' In B. Gordijn & A. M. Cutter (Eds.), *In pursuit of nanoethics* (pp. 31–38). Springer.

Holm, S., & McNamee, M. (2011). Physical enhancement: What baseline, whose judgment? In J. Savulescu, R. ter Meulen, & G. Kahane (Eds.), *Enhancing human capacity* (pp. 291–303). Wiley-Blackwell.

Hook, C. C. (2007). Nanotechnology and the future of medicine. In N. Cameron & E. Mitchell (Eds.), *Nanoscale: Issues and perspectives for the nano century* (pp. 337–357). Wiley.

Hope, T. (2011). Cognitive therapy and positive psychology combined: A promising approach to the enhancement of happiness. In J. Savulescu, R. ter Meulen, & G. Kahane (Eds.), *Enhancing human capacity* (pp. 230–244). Wiley-Blackwell.

Hopkins, P. D. (2008). A moral vision of transhumanism. *Journal of Evolution and Technology, 19*, 3–7.

Horrobin, S. (2011). The value of life extension to persons as conatively driven processes. In J. Savulescu, R. ter Meulen, & G. Kahane (Eds.), *Enhancing human capacity* (pp. 421–434). Wiley-Blackwell.

Hottois, G. (2018). *Philosophie et ideologies trans/posthumanistes*. Vrin.

Housden, C. R., Morein-Zamir, S., & Sahakian, B. J. (2011). Cognitive enhancing drugs: Neuroscience and society. In J. Savulescu, R. ter Meulen, & G. Kahane (Eds.), *Enhancing human capacity* (pp. 113–126). Wiley-Blackwell.

Howse, K. (2011). Policy-Making for a new generation of interventions in age-related disease and decline. In J. Savulescu, R. ter Meulen, & G. Kahane (Eds.), *Enhancing human capacity* (pp. 453–463). Wiley-Blackwell.

Hughes, J. (2004). *Citizen cyborg: Why democratic societies must respond to the redesigned human of the future*. Westview Press Books.

Hughes, J. (2010). Technoprogressive biopolitics and human enhancement. In J. D. Moreno & S. Berger (Eds.), *Progress in Bioethics: Science, policy, and politics* (pp. 163–188). MIT Press.

Hughes, J. (2001). The future of death: Cryonics and the telos of liberal individualism. *Journal of Evolution and Technology, 6*(1), 1–30.

Hughes, J. (2014). Politics. In R. Ranisch & S. L. Sorgner (Eds.), *Post- and transhumanism: An introduction* (pp. 133–148). Peter Lang.

Hume, D. (2004). *A treatise of human nature*. Digireads Publishing.

Huxley, A. (1932). *Brave new world*. http://www.huxley.net/bnw/1st-edition.html

Huxley, J. (1927). *Religion without revelation*. Harper and Brothers.

Huxley, J. (1957). Transhumanism. In *New bottles for new wine*. Chatto & Windus.

Illes, J. (2006). *Neuroethics: Defining the issues in theory, practice, and policy*. Oxford University Press.

Illich, I. (1975). *Medical nemesis: The expropriation of health*. Pantheon Books.

Israel, G. (2004). *La macchina vivente. Contro le visioni meccanicistiche dell'uomo*. Bollati-Boringhieri.

Istvan, Z. (2013). *The transhumanist wager*. Futurity Imagine Media.

Iuculano, T., & Kadosh, R. C. (2013). the mental cost of cognitive enhancement. *Journal of Neuroscience, 33*(10), 4482–4486.

Jacobson, N. (2012). *Dignity and health*. Vanderbilt University Press.

Jonas, H. (1962/2001). Immortality and the modern temper. In *The Phenomenon of life* (pp. 262–281). Northwestern University Press.

Jonas, H. (1996). The burden and blessing of mortality. In *Mortality and morality: A search for the good after Auschwitz* (pp. 87–98). Northwestern University Press.

Jones, G., & Whitaker, M. (2012). Transforming the human body. In C. Blake, C. Molloy & S. Shakespeare (Eds.), *Beyond human: From animality to transhumanism* (pp. 254–279). Bloomsbury.

Jotterand, F. (2008). Beyond therapy and enhancement: The alteration of human nature. *NanoEthics, 21*, 15–23.

Jotterand, F. (2010). Human dignity and transhumanism: Do anthro-technological devices have moral status? *The American Journal of Bioethics, 10*(7), 45–52.

Joy, B. (2000). Why the future doesn't need us. In F. Allhoff, et al. (Eds.), *Nanoethics: The ethical and social implications of nanotechnology* (pp. 17–30). Wiley.

Juengst, E. T. (1997). Can enhancement be distinguished from prevention in genetic medicine? *Journal of Medicine and Philosophy, 22*(2), 125–142.

Juengst, E. T. (1998). What does enhancement mean. In E. Parens (Ed.), *Enhancing human traits* (pp. 29–47). Georgetown University Press.

Juengst, E. T. (2009). What's taxonomy got to do with it? 'Species integrity', human rights, and science policy. In J. Savulescu & N. Bostrom, *Human enhancement* (pp. 43–58). Oxford University Press.

Juth, N. (2011). Enhancement, autonomy, and authenticity. In J. Savulescu, R. ter Meulen, & G. Kahane (Eds.), *Enhancing human capacity* (pp. 34–48). Wiley-Blackwell.

Kahane, G. (2001). Reasons to feel, reasons to take pills. In J. Savulescu, R. ter Meulen, & G. Kahane (Eds.), *Enhancing human capacity* (pp. 166–178). Wiley-Blackwell.

Kamm, F. M. (2009). What is and is not wrong with enhancement? In J. Savulescu & N. Bostrom (Eds.), *Human enhancement* (pp. 91–130). Oxford University Press.

Kant, I. (1991). The contest of faculties: A renewed attempt to answer the question: 'Is the human race continually improving?'. In *Political writings* (pp. 176–190). Cambridge University Press.

Kant, I. (2006). *Anthropology from a pragmatic point of view*. Cambridge University Press.

Kass, L. R. (1985). *Toward a more natural science: Biology and human affairs*. The Free Press.

Kass, L. R. (1997). Wisdom of repugnance: Why we should ban the cloning of humans. *The Valparaiso University Law Review, 32*, 679.

Kass, L. R. (2002). *Life, liberty and the defense of dignity: The challenge for bioethics*. Encounteurs Book.

Katz, B. F. (2008). *Neuroengineering the future: Virtual minds and the creation of immortality*. Infinity Science Press.
Kearnes, M. B., Macnaghten, P. M., & Wilsdon, J. (2006). *Governing at the nanoscale: People, policies and emerging technologies*. DEMOS.
Keiper, A. (2007). Nanoethics as a discipline. *The New Atlantis, 16*, 55–67.
Kekich, D. (2010). *Life extension express*. Max Life Foundation.
Kelly, K. (2006). *The inevitable: Understanding the 12 technological forces that will shape our future*. Viking.
Kelly, K. (2010). *What technology wants*. Viking.
Khushf, G. (2004). Systems theory and the ethics of human enhancement: A framework for NBIC convergence. *Annals of the New York Academy of Sciences, 1013*(1), 124–149.
Khushf, G. (2005). The use of emergent technologies for enhancing human performance: Are we prepared to address the ethical and policy issues. *Public Policy and Practice, 4*(2), 1–17.
Kirkwood, T., Bond, J., May, C., McKeith, I., & The, M.-M. (2009). Mental capital through life: Future challenges. In C. Cooper, U. Goswami, & B. J. Sahakian (Eds.), *Mental capital and well-being* (3–53). Wiley.
Knorr, C. K. (2005). The rise of a culture of life. *EMBO Reports, 6*, 77–80.
Kolnai, A. (1976). Dignity. *Philosophy, 51*, 251–271.
Koops, B. J., Lüthy, C. H., Nelis, A., Sieburgh, C., Jansen, J. P. M., & Schmid, M. S. (2013). *Engineering the human: Human enhancement between fiction and fascination*. Springer.
Kosfeld, M., Heinrichs, M., Zak, P. J., Fischbacher, U., & Fehr, E. (2005). Oxytocin increases trust in humans. *Nature Cell Biology, 435*(7042), 673–676.
Kramer, P. D. (1993). *Listening to Prozac*. Penguin.
Kranzberg, M. (1986). Technology and history: 'Kranzberg's Laws.' *Technology and Culture, 27*, 544–560.
Kroes, P., & Verbeek, P.-P. (2014). *The moral status of technical artefacts*. Springer.
Kuhn, T. S. (2012). *The structure of scientific revolutions* (4th ed.). University of Chicago Press.
Kurzweil, R. (1999). *The age of spiritual machines*. Penguin.
Kurzweil, R. (2005). *The singularity is near*. Penguin.
Kurzweil, R. (2012). *How to create a mind: The secret of human thought revealed*. Pengin.
Kurzweil, R., & Grossman, T. (2004a). *Fantastic voyage: Live long enough to live forever*. Rodale Books.

Kurzweil, R., & Grossman, T. (2004b). *Transcend: Nine steps to living well forever*. Rodale Books.

Landeweerd, L. (2011). Asperger's syndrome, bipolar disorder and the relation between mood, cognition, and well-being. In J. Savulescu, R. ter Meulen, & G. Kahane (Eds.), *Enhancing human capacity* (pp. 207–217). Wiley-Blackwell.

Lanigan, B. T. (2009). *Human dignity and bioethics*. Nova Science.

Lavazza, A., & Reichlin, M. (2019). Moral enhancement. *Topoi, 38*(1).

Le Devedec, N. (2015). *La société de l'amélioration. La perfectibilité humaine des Lumières au transhumanisme*. Liber.

LeDoux, J. (2003). *Synaptic self: How our brains become who we are*. Penguin.

Lee, N. (2019a). *The transhumanism handbook*. Springer.

Lee, N. (2019b). Brave new world of transhumanism. In N. Lee (Ed.), *The transhumanism handbook* (pp. 3–48). Springer.

Leibniz, G. W. (2005). *Confessio Philosophi: Papers concerning the problem of evil,* 1671–1678. Yale University Press.

Levin, Y. (2003). The paradox of conservative bioethics. *New Atlantis, 1*(Spring), 53–65.

Levy, N. (2007). *Neuroethics: Challenges for the 21st century*. Cambridge University Press.

Lewis, S. C. (1943). *The abolition of man*. Samizdat University Press.

Liao, S. M., & Roache, R. (2011). After Prozac. In J. Savulescu, R. ter Meulen, & G. Kahane (Eds.), *Enhancing human capacity* (pp. 245–256). Wiley-Blackwell.

Liao, S. M., & Sandberg, A. (2008). The normativity of memory modification. *Neuroethics, 12*, 85–99.

Lilley, S. (2013). *Transhumanism and society: The social debate over human enhancement*. Springer.

Lin, P. (2007). Nanotechnology bound: Evaluating the case for more regulation. *NanoEthics, 1*(2), 105–122.

Lin, P., Abney, K., & Bekey, G. A. (2012). *Robot ethics: The ethical and social implications of robotics*. MIT Press.

Lin, P., & Allhoff, F. (2006). Nanoethics and human enhancement: A critical evaluation of recent arguments. *Nanotechnology Perceptions, 2*(1), 47.

Lin, P., & Allhoff, F. (2008a). Against unrestricted human enhancement. *Journal of Evolution and Technology, 18*(1), 35.

Lin, P., & Allhoff, F. (2008b). Untangling the debate: The ethics of human enhancement. *Nanoethics, 2*(3), 251–264.

Locke, J. (1768). *An essay concerning human understanding*. H. Woodfall.

Loland, S. (2011). Can a ban on doping in sport be morally justified? In J. Savulescu, R. ter Meulen, & G. Kahane (Eds.), *Enhancing human capacity* (pp. 326–311). Wiley-Blackwell.

Lucivero, F., & Vedder, A. (2013). *Beyond therapy vs. enhancement*. Pisa University Press.

Lynch, Z. (2004). Neurotechnology and society 2010–2060. *Annals of the New York Academy of Sciences, 1013*(1), 229–233.

Lynch, Z. (2005). Neuropolicy 2005–2035: Converging technologies enable neurotechnology creating new ethical dilemmas. In W. S. Bainbridge & M. C. Roco (Eds.), *Managing Nano-Bio-Info-Cogno innovations: Converging technologies in society* (pp. 173–191). Springer.

MacKenzie, D. A., & Wajcman, J. (1985). *The social shaping of technology: How the refrigerator got its hum*. Open University Press.

Macklin, R. (2003). Dignity is a useless concept. *British Medical Journal, 327*(7429), 1419–1420.

Macklin, R. (2006). The new conservatives in bioethics: Who are they and what do they seek? *Hastings Center Report, 361*, 34–43.

Mader, K., Brune, H., Ernst, H., Grunwald, A., Grünwald, W., Hofmann, H., et al. (2006). *Nanotechnology: Assessment and perspectives*. Springer Verlag.

Marcus, S. (2002). *Neuroethics: Mapping the fields*. Dana Press.

Marzocco, R. (2019). *Transhumanism. Engineering the human condition: History, philosophy and current status*. Springer.

Maslow, A. H. (1998). *Toward a psychology of being*. Wiley.

Mayer, C. (2011). *Amortality: The pleasures and perils of living agelessly*. Random House.

McKenny, G. P. (1997). *To relieve the human condition: Bioethics, technology, and the body*. State University of New York Press.

Mečnikov, I. (1903). *Études sur la nature humaine. Essais de philosophie optimiste*. Masson.

Mehlman, M. J. (2000). The law of above averages: Leveling the new genetic enhancement playing field. *Iowa Law Review, 85*, 517–593.

Mehlman, M. J. (2003). *Wondergenes: Genetic enhancement and the future of society*. Indiana University Press.

Mehlman, M. J. (2009). *The price of perfection: Individualism and society in the era of biomedical enhancement*. Johns Hopkins University Press.

Merleau-Ponty, M. (2014). *Phenomenology of perception*. Routledge.

Metalnikov, S. (1937). *La lutte contre la mort*. Gallimard.

Miah, A. (2003). Be very afraid: Cyborg athletes, transhuman ideals and posthumanity. *Journal of Evolution and Technology, 13*(2), 1–23.

Miah, A. (2004). *Genetically modified athletes: Biomedical ethics, gene doping and sport*. Routledge.
Miah, A. (2008). Justifying human enhancement: The accumulation of biocultural capital. In *Perspectives on our ethical future: Boundaries to human enhancements*. Royal Society for the Encouragement of Arts, Manufactures and Commerce.
Miah, A. (2009). A Critical History of Posthumanism. In A. M. Cutter, B. Gordijn, G. E. Marchant, A. Pompidou, & R. Chadwick (Eds.), *Medical Enhancement and posthumanity* (pp. 71–94). Springer.
Miah, A. (2011). Physical enhancement: The state of the art. In J. Savulescu, R. ter Meulen, & G. Kahane (Eds.), *Enhancing human capacity* (pp. 266–273). Wiley-Blackwell.
Milburn, C. (2002). Nanotechnology in the age of posthuman engineering: Science fiction as science. *Configurations, 10*(2), 261–295.
Mill, J. S. (1863). *On liberty*. Ticknor and Fields.
Miller, P., Parker, S., & Gillinson, S. (2004). *Disablism: How to tackle the last prejudice*. DEMOS.
Miller, P., & Wilsdon, J. (2006). *Better humans? The politics of human enhancement and life extension*. DEMOS.
Minsky, M. (1986). *The society of mind*. Simon & Schuster.
Minsky, M. (2006). *The emotion machine: Commonsense thinking, artificial intelligence, and the future of the human mind*. Simon & Schuster.
Missa, J.-N., & Perbal, L. (2009). *"Enhancement": Ethique et philosophie de la médecine d'amélioration*. Vrin.
Mitcham, C. (2014). Agency in humans and in artifacts: A contested discourse. In P. Kroes & P.-P. Verbeek (Eds.), *The moral status of technical artefacts* (pp. 11–30). Springer.
Mlle de Saint-Quentin. (1692–1693). Traité pour établir la possibilité de l'immortalité corporelle. *Mercure Galant, 33–34*, 120–124.
Moatti, A. (2020). *Aux racines du transhumanisme: France, 1930–1980*. Odile Jacob.
Molloy, C., Black, C., & Shakespeare, S. (2012). *Beyond human: From animality to transhumanism*. Continuum.
Moore, P. (2008). *Enhancing me: The hope and the hype of human enhancement*. Wiley.
Moravec, H. (1988). *Mind children: The future of robot and human intelligence*. Harvard University Press.

Mordacci, R. (2011). Intergenerational justice and lifespan extension. In J. Savulescu, R. ter Meulen, & G. Kahane (Eds.), *Enhancing human capacity* (pp. 410–420). Wiley-Blackwell.

More, M. (1995). *The diachronic self: Identity, continuity, transformation*. A Bell and Howell Company.

More, M. (1996). *Transhumanism: Towards a futurist philosophy*. https://web.archive.org/web/20051029125153/http://www.maxmore.com/transhum.htm. Last accessed 26 April 2020.

More, M. (2010). The overhuman in the transhuman. *Journal of Evolution and Technology, 21*(1), 1–4.

More, M. (2011). True transhumanism: A reply to Don Ihde. In G. R. Hansell & W. J. Grassie (Eds.), *H+ transhumanism and its critics* (pp. 62–66). Metanexus Institute.

More, M. (2013a). The philosophy of transhumanism. In M. More & N. Vita-More (Eds.), *The transhumanist reader* (pp. 3–17). Wiley-Blackwell.

More, M. (2013b). The proactionary principle: Optimizing technological outcomes. In M. More & N. Vita-More (Eds.), *The transhumanist reader* (pp. 258–267). Wiley-Blackwell.

More, M. (2013c). A letter to mother nature. In M. More & N. Vita-More (Eds.), *The transhumanist reader* (pp. 449–450). Wiley-Blackwell.

More, M., & Vita-More, N. (2013). *The transhumanist reader*. Wiley Blackwell.

Mulhall, D. (2002). *Our molecular future: How nanotechnology, robotics, genetics and artificial intelligence will transform our world*. Prometheus Books.

Naam, R. (2005). *More than human*. Broadway Books.

Nayar, P. K. (2013). *Posthumanism*. Polity Press.

Neiman, S. (2014). *Why grow up? Subversive thoughts for an infantile age*. Penguin Books.

Nichter, M. (2008). *Global health: Why cultural perceptions, social representations, and biopolitics matter*. University of Arizona Press.

Nisbet, J. (1994). *History of the idea of progress* (II). Transaction Publisher.

Niyadurupola, G. (2008). Better brains. In L. Zonneveld (Ed.), *Reshaping the human condition: Exploring human enhancement* (pp. 77–88). Rathenau Institute.

Nordenfelt, L. (2003). An evolutionary concept of health: Health as a natural function. In L. Nordenfelt & P.-E. Liss (Eds.), *Dimensions of Health and Health Promotion* (pp. 37–54). Editions Rodopi.

Nussbaum, M. C. (2001). *Upheavals of thought: The intelligence of emotions*. Cambridge University Press.

Nussbaum, M., & Sen, A. (1993). *The quality of life*. Clarendon Press.

O'Connell, M. (2017). *To be a machine: Adventures among cyborgs, utopians, hackers and the futurists*. Doubleday.

Olivier, P., Wherton, J., & Monk, A. (2009). Technology solutions to prevent waste of mental capital. In C. Cooper, U. Goswami, & B. J. Sahakian (Eds.), *Mental capital and well-being* (pp. 269–274). Wiley.

Olsen, J. K. B., Pedersen, S. A., & Hendricks, V. F. (2009). *A companion to the philosophy of technology*. Wiley-Blackwell.

Olson, E. T. (1997). *The human animal: Personal identity without psychology*. Oxford University Press.

O'Mathuna, D. (2009). *Nanoethics: Big ethical issues with small technology*. Continuum.

Overall, C. (2003). *Aging, death and human longevity*. University of California Press.

Overall, C. (2011). Lifespan extension: Metaphysical basis and ethical outcomes. In J. Savulescu, R. ter Meulen, & G. Kahane (Eds.), *Enhancing human capacity* (pp. 386–397). Wiley-Blackwell.

Parens, E. (1998a). *Enhancing human Ttraits: Ethical and social implications*. Georgetown University Press.

Parens, E. (1998b). Is better always good? The enhancement project. In E. Parens (Ed.), *Enhancing human traits: Ethical and social implications* (pp. 1–28). Georgetown University Press.

Parens, E. (2005). Authenticity and ambivalence: Toward understanding the enhancement debate. *The Hastings Center Report, 35*(3), 34–41.

Parens, E. (2006). Creativity, gratitude, and the enhancement debate. In J. Illes (Ed.), *Neuroethics: Defining the issues in theory, practice and policy* (pp. 75–86). Oxford University Press.

Parens, E. (2013). On good and bad forms of medicalization. *Bioethics, 27*(1), 28–35.

Parfit, D. (1971). Personal identity. *The Philosophical Review, 80*(1), 3–27.

Passmore, J. (2000). *The perfectibility of man*. Liberty Fund.

Paul, G. S., & Cox, E. (1996). *Beyond humanity: Cyberevolution and future minds*. Charles River.

PCB. (2003). *Beyond therapy: Biotechnology and the pursuit of happiness*. President's Council on Bioethics. Harper & Collins.

Pearce, D. (1995). *The hedonistic imperative*. www.hedweb.com

Pence, G. E. (2012). *How to build a better human: An ethical blueprint*. Rowman & Littlefield Publishers.

Pepperell, R. (2004). *The posthuman condition: Consciousness beyond the brain*. Intellect Books.

Pepperell, R. (2005). Posthumans and extended experience. *Journal of Evolution and Technology, 14*(1), 27–41.

Persson, I., & Savulescu, J. (2008). The perils of cognitive enhancement and the urgent imperative to enhance the moral character of humanity. *Journal of Applied Philosophy, 25*(3), 162–177.

Persson, I., & Savulescu, J. (2010). Moral transhumanism. *Journal of Medicine and Philosophy, 35*(6), 656–669.

Persson, I., & Savulescu, J. (2011a). The turn for ultimate harm: A reply to Fenton. *Journal of Medical Ethics, 37*(7), 441–444.

Persson, I., & Savulescu, J. (2011b). Unfit for the future? Human nature, scientific progress, and the need for moral enhancement. In J. Savulescu, R. ter Meulen, & G. Kahane (Eds.), *Enhancing human capacity* (pp. 486–499). Wiley-Blackwell.

Persson, I., & Savulescu, J. (2012a). *Unfit for the future: The need for moral enhancement*. Oxford University Press.

Persson, I., & Savulescu, J. (2012b). Moral enhancement, freedom and the god machine. *The Monist, 95*(3), 399–442.

Persson, I., & Savulescu, J. (2014). Should moral bioenhancement be compulsory? Reply to Vojin Rakic. *Journal of Medical Ethics, 40*(4), 251–252.

Pessina, A. (2016). *L'Io insoddisfatto. Tra Prometeo e Dio.* Vita & Pensiero.

Pessina, A. (2019). A clay person: The promises of moral bioenhancement. In A. Lavazza & M. Reichlin (Eds.), Moral enhancement. *Topoi, 38*(1), 87–93.

Peters, T. (2003). *Playing God: Genetic determinism and human freedom.* Routledge.

Peters, T. (2007). Are we playing god with nanoenhancements? In F. Allhoff, P. Lin, J. Moor, & J. Weckert (Eds.), *Nanoethics: The ethical and social implications of nanotechnology* (pp. 173–183). Wiley.

Pico della Mirandola. (1985). *On the dignity of man, on being and the one, Heptaplus.* Macmillan.

Pinker, S. (2008, May 8). The stupidity of dignity. *The New Republic*. http://www.newrepublic.com/article/the-stupidity-dignity. Last accessed 11 June 2020.

Post, S. G., & Binstock, R. H. (2004). *The fountain of youth.* Oxford University Press.

Powell, R., & Buchanan, A. (2011). Breaking evolution's chains: The promise of enhancement by design. In J. Savulescu, R. ter Meulen, & G. Kahane (Eds.), *Enhancing human capacities* (pp. 49–67). Wiley-Blackwell.

Rabinow, P., & Rose, N. (2006). Biopower today. *BioSocieties, 1*(2), 195–217.

Racine, E. (2010). *Pragmatic neuroethics: Improving treatment and understanding of the mind-brain.* MIT Press.

Racine, E., & Forlini, C. (2010). Cognitive enhancement, lifestyle choice or misuse of prescription drugs? *Neuroethics, 3*(1), 1–4.

Rakić, V. (2014). Voluntary moral enhancement and the survival-at-any-cost bias. *Journal of Medical Ethics, 40*, 246–250.

Ramsey, P. (1977). *Fabricated man: The ethics of genetic control*. Yale University Press.

Rand, A. (1961). *The virtue of selfishness: A new concept of egoism.* New American Library.

Ranisch, R., & Sorgner, S. L. (2014). *Post- and transhumanism: An introduction.* Peter Lang.

Ratner, M. A., & Ratner, D. (2003). *Nanotechnology: A gentle introduction to the next big idea.* Prentice Hall Professional.

Ravelingien, A., Braeckman, J., Crevits, L., De Ridder, D., & Mortier, E. (2009). 'Cosmetic neurology' and the moral complicity argument. *Neuroethics, 2*(3), 151–162.

Rawls, J. (1971). *A theory of justice*. Harvard University Press.

Rawls, J. (2001). *Justice as fairness* (2nd ed.). Harvard University Press.

Regis, E. (1990). *Great Mambo Chicken and the transhuman condition.* Penguin Books.

Reichlin, M. (2019). The moral agency argument against moral bioenhancement. In A. Lavazza & M. Reichlin (Eds.), Moral enhancement. *Topoi, 38*(1), 53–62.

Reiner, P. (2013). The biopolitics of cognitive enhancement. In E. Hildt & A. Franke (Eds.), *Cognitive enhancement: An interdisciplinary perspective* (pp. 189–200). Springer.

Repantis, D., Schlattmann, P., Laisney, O., & Heuser, I. (2010). Modafinil and methylphenidate for neuroenhancement in healthy individuals: A systematic review. *Pharmacological Research, 62*(3), 187–206.

Roache, R. (2008). Ethics, speculation, and values. *NanoEthics, 2*(3), 317–327.

Roache, R., & Clarke, S. (2009). Bioconservatism, bioliberalism, and the wisdom of reflecting on repugnance. *Monash Bioethics Review, 28*(4), 1–21.

Roco, M. C., & Bainbridge, W. S. (2002). *Converging technologies for improving human performance: Nanotechnology, biotechnology, information technology and cognitive science*. NSF/Department of Commerce Arlington.

Roco, M. C., & Bainbridge, W. S. (2007). *Nanotechnology: Societal implications II—Individual perspectives*. Springer.

Roco, M. C., Bainbridge, W. S., Tonn, B., & Whitesides, G. (2013). *Convergence of knowledge, technology and society: Beyond convergence of Nano-bio-info-cognitive technologies*. Springer.

Roco, M. C., Williams, R. S., & Alivisatos, P. (2000). *National research directions*. Kluwer Academic Publishers.

Rose, N. (2007). *The politics of life itself: Biomedicine, Power, and subjectivity in the twenty-first century*. Princeton University Press.

Rose, S. (2006). Brain Gain. In P. Miller & J. Wilsdon (Eds.), *Better humans? The politics of human enhancement and life extension* (pp. 69–78). DEMOS.

Rose, M. R., et al. (2004). *The scientific conquest of death: Essays on infinite lifespans*. Libros en Red.

Roskies, A. (2010). Neuroethics: Considering its scope and limits. *AJOB Neuroscience, 1*(4), 1–2.

Rothman, S., & Rothman, D. (2003). *The pursuit of perfection: The promise and perils of medical enhancement*. Pantheon Books.

Rousseau, J.-J. (1755/1997). *Discours on the origin and foundation of inequality among men or second discourse*. In *The discourse and other early political writings* (pp. 111–222). Cambridge University Press.

Rubin, C. (2009). Commentary on Bostrom. In B. T. Lanigan (Ed.), *Human dignity and bioethics* (pp. 145–147). Nova Science.

Sandberg, A. (n.d.). *We, borg: Speculations on hive minds as a posthuman state*. http://www.aleph.se/Trans/Global/Posthumanity/WeBorg.html. Last accessed 19 March 2020.

Sandberg, A. (2003). Morphological freedom: Why we not just want it, but need it. In M. More & N. Vita-More (Eds.), *The transhumanist reader* (pp. 56–64). Wiley Blackwell.

Sandberg, A. (2011). Cognition enhancement: Upgrading the brain. In J. Savulescu, R. ter Meulen, & G. Kahane (Eds.), *Enhancing human capacity* (pp. 71–91). Wiley-Blackwell.

Sandberg, A., & Bostrom, N. (2008). *Whole brain emulation: A roadmap*. Future of Humanity Institute, Oxford University. http://www.fhi.ox.ac.uk/reports/2008-3.pdf

Sandberg, A., & Savulescu, J. (2011). The social and economic impacts of cognitive enhancement. In J. Savulescu, R. ter Meulen, & G. Kahane (Eds.), *Enhancing human capacity* (pp. 92–112). Wiley-Blackwell.

Sandel, M. (2004). The case against perfection. *The Atlantic Monthly*, 1–10.

Sandler, R. L. (2014). *Ethics and emerging technologies*. Palgrave Macmillan.

Satz, A., & Silvers, A. (2000). Disability and biotechnology. In T. Murray (Ed.), *Encyclopaedia of ethical, legal, and policy issues in biotechnology* (pp. 173–187). Wiley-Interscience.

Savulescu, J. (2001). Procreative beneficence: Why we should select the best children. *Bioethics, 15*(5/6), 413–426.

Savulescu, J. (2002). Deaf Lesbians, 'Designer Disability', and the future of medicine. *BMJ, 325*(7367), 771–773.

Savulescu, J. (2006). Justice, fairness, and enhancement. *Annals of the New York Academy of Sciences, 1093*(1), 321–338.

Savulescu, J. (2007). Genetic interventions and the ethics of enhancement of human beings. In B. Steinbock (Ed.), *The Oxford handbook of bioethics* (pp. 516–535). Oxford University Press.

Savulescu, J. (2009). The human prejudice and the moral status of enhanced beings: What do we owe the gods? In N. Bostrom & J. Savulescu (Eds.), *Human enhancement* (pp. 211–247). Oxford University Press.

Savulescu, J., Bennett, F., & Clayton, M. (2004). Why we should allow performance enhancing drugs in sport. *British Journal of Sports Medicine, 38*, 666–670.

Savulescu, J., & Bostrom, N. (2009). *Human enhancement*. Oxford University Press.

Savulescu, J., & Foddy, B. (2011). Le Tour and failure of zero tolerance: Time to relax doping controls. In J. Savulescu, R. ter Meulen, & G. Kahane (Eds.), *Enhancing human capacities* (pp. 304–312). Wiley-Blackwell.

Savulescu, J., Sandberg, A., & Kahane, G. (2011). Well-Being and enhancement. In J. Savulescu, R. ter Meulen, & G. Kahane (Eds.), *Enhancing human capacities* (pp. 3–18). Wiley-Blackwell.

Savulescu, J., ter Meulen, R., & Kahane, G. (2011). *Enhancing human capacities*. Wiley-Blackwell.

Schechtman, M. (2007). *The constitution of selves*. Cornell University Press.

Schechtman, M. (2014). *Staying Alive: Personal identity, practical concerns, and the unity of a life*. Oxford University Press.

Schermer, M., & Bolt, I. (2011). What's in a name? ADHD and the Gray area between treatment and enhancement. In J. Savulescu, R. ter Meulen, & G. Kahane (Eds.), *Enhancing human capacities* (pp. 179–193). Wiley-Blackwell.

Schopenhauer, A. (1969). *The world as will and representation*. Dover.

Searle, J. (1995). *The construction of social reality*. Free Press.

Selgelid, M. J. (2009). Afterword: Advancing posthuman enhancement dialogue. In B. Gordijn & R. Chadwick (Eds.), *Medical enhancement and posthumanity* (pp. 237–240). Springer.

Shakespeare T. 2014. Foreword: Five thoughts about enhancement. In M. Eilers, K. Grüber, & C. Rehmann-Sutter (Eds.), *The human enhancement debate and disability: New bodies for a better life* (pp. ix–xiii). Palgrave Macmillan.

Sharon, T. (2012). A cartography of the posthuman humanist, non-humanist and mediated perspectives on emerging biotechnologies. *Krisis, 2*, 4–19.

Short, S. (2005). *Cyborg cinema and contemporary subjectivity*. Palgrave Macmillan.

Shostak, S. (2002). *Becoming immortal*. State University of New York Press.

Simon, Z. B. (2019). Two cultures of the posthuman future. *History and Theory, 58*(2), 171–184.

Simondon, G. (2001). *Du mode d'existence des objets techniques*. Aubier.

Smith, M., & Morra, J. (2006). *The prosthetic impulse: From a posthuman present to a biocultural future*. The MIT Press.

Soper, K. (1986). *Humanism and anti-humanism*. Hutchinson.

Sorgner, S. L. (2009). Nietzsche, the overhuman, and transhumanism. *Journal of Evolution and Technology, 20*(1), 29–42.

Sorgner, S. L. (2013a). Evolution, education, and genetic enhancement. In S. L. Sorgner, B.-R. Jovanovic, & N. Grimm (Eds.), *Evolution and the future: Anthropology, ethics, religion* (pp. 85–100). Peter Lang.

Sorgner, S. L. (2013b). Human dignity 2.0: Beyond a rigid version of anthropocentrism. *Trans-Humanities, 61*, 135–159.

Sorgner, S. L., & Grimm, N. (2013). Introduction: Evolution today. In S. L. Sorgner, B.-R. Jovanovic & N. Grimm (Eds.), *Evolution and the future: Anthropology, ethics, religion* (pp. 9–20). Peter Lang.

Sorgner, S. L., Jovanovic, B.-R., & Grimm, N. (2013). *Evolution and the future: Anthropology, ethics, religion*. Peter Lang.

Sparrow, R. (2011). A not-so-new eugenics: Harris and Savulescu on human enhancement. *Hastings Center Report, 41*(1), 32–42.

Sparrow, R. (2014). Better living through chemistry? A reply to Savulescu and Persson on 'Moral Enhancement.' *Journal of Applied Philosophy, 31*(4), 23–32.

Spinoza, B. (1985). *Ethics*. In *The collected writings of Spinoza* (2 vols., Vol. 1). Princeton University Press.

Staman, J., Dijstelbloem, H., & Smits, M. (2008). Afterword. In L. Zonneveld (Ed.), *Reshaping the human condition: Exploring human enhancement* (pp. 149–157). Rathenau Institute.

Stock, G. (1993). *Metaman: The merging of humans and machines into a global superorganism*. Transworld.

Stock, G. (2003). *Redesigning humans*. Profile Books.
Taguieff, P.-A. (2001). *Du progres*. J'ai lu.
Tamburrini, C., & Tännsjö, T. (2011). Enhanced bodies. In J. Savulescu, R. ter Meulen, & G. Kahane (Eds.), *Enhancing human capacity* (pp. 274–290). Wiley-Blackwell.
Taylor, C. (1991). *The ethics of authenticity*. Harvard University Press.
Taylor, J. S. (2008). *Personal autonomy: New essays on personal autonomy and its role in contemporary moral philosophy*. Cambridge University Press.
Teilhard de Chardin, P. (1959). *L'Avenir de L'Homme*. Editions de Seuil.
Temkin, L. (2011). Is living longer living better? In J. Savulescu, R. ter Meulen, & G. Kahane (Eds.), *Enhancing human capacity* (pp. 350–367). Wiley-Blackwell.
ter Meulen, R. (2015). The moral ambiguity of human enhancement. In S. Bateman, J. Gayon, S. Allouche, J. Goffette, & M. Marzano (Eds.), *Inquiring into human enhancement: Interdisciplinary and International Perspectives* (pp. 86–99). Palgrave Macmillan.
Tirosh-Samuelson, H. (2011a). Science and the betterment of humanity: Three British prophets of transhumanism. In H. Tirosh-Samuelson & K. L. Mossman (Eds.), *Building better humans?* (pp. 55–82). Peter Lang.
Tirosh-Samuelson, H. (2011b). Engaging transhumanism. In G. R. Hansell & W. Grassie (Eds.), *H±: Transhumanism and its critics* (pp. 10–20). Metanexus Institute.
Tirosh-Samuelson, H. (2014). Religion. In R. Ranisch & S. L. Sorgner (Eds.), *Post- and transhumanism: An introduction* (pp. 49–71). Peter Lang.
Tirosh-Samuelson, H., & Mossman, K. L. (2011a). *Building better Humans?* Peter Lang.
Tirosh-Samuelson, H., & Mossman, K. L. (2011b). New perspective on transhumanism. In H. Tirosh-Samuelson & K. L. Mossman (Eds.), *Building better Humans?* (pp. 29–52). Peter Lang.
Toffoletti, K. (2007). *Cyborgs and Barbie dolls: Feminism, popular culture and the posthuman body*. Tauris.
Turner, D., & Sahakian, B. (2006). Neuroethics of cognitive enhancement. *BioSocieties, 1*(1), 113–123.
Turner, D., & Sahakian, B. (2008). The cognition-enhanced classroom. In L. Zonneveld, H. Dijstelbloem, & D. Ringoir (Eds.), *Reshaping the human condition: Exploring human enhancement* (pp. 107–114). Rathenau Institute.
Verbeek, P.-P. (2011). *Moralizing technology: Understanding and designing the morality of things*. University of Chicago Press.

Verdoux, P. (2009). Transhumanism, progress and the future. *Journal of Evolution and Technology, 20*(2), 49–69.

Vita-More, N. (2009). Introduction to "H+: Transhumanism Answers Its Critics". *Metanexus*. https://metanexus.net/introduction-h-transhumanism-answers-its-critics/. Last accessed 26 April 2020.

Vita-More, N. (2013). Aesthetics: Bringing the arts & design into the discussion of transhumanism. In M. More & N. Vita-More (Eds.), *The transhumanist reader* (pp. 18–27). Wiley Blackwell.

Vita-More, N. (2019a). History of transhumanism. In N. Lee (Ed.), *The transhumanism handbook* (pp. 49–61). Springer.

Vita-More, N. (2019b). *Transhumanism: What is it?* Bowker.

Vos, R. (2011). What is good or bad in mood enhancement? In J. Savulescu, R. ter Meulen, & G. Kahane (Eds.), *Enhancing human capacity* (pp. 194–206). Wiley-Blackwell.

Walker, M. (2008). Cognitive enhancement and the identity objection. *Journal of Evolution and Technology, 18*(1), 108–115.

Walker, M. (2011). Ship of fools: Why transhumanism is the best bet to prevent the extinction of civilization. In G. R. Hansell & W. J. Grassie (Eds.), *H+ transhumanism and its critics* (pp. 43–50). Metanexus Institute.

Walker, M. (2014). *Transhumanism and Radical Enhancement*. http://eugenicsarchive.ca/discover/tree/545f9d2dbb4dbcc59800000. Last accessed 26 April 2020.

Warwick, K. (2004). *I, Cyborg*. University of Illinois Press.

Warwick, K. (2014). The cyborg revolution. *Nanoethics, 8*(3), 263–273.

Watson, G. (1982). *Free Will*. Oxford University Press.

Weckert, J. (2016). Playing God: What is the problem? In S. Clarke, J. Savulescu, T. Coady, A. Giubilini, & S. Sanyal (Eds.), *The ethics of human enhancement: Understanding the debate* (pp. 87–99). Oxford University Press.

Weismann, A. (1892). *Essais sur l'hérédité et la sélection naturelle*. Reinwald.

WHO. (1946). *Constitution of the World Health Organization*. World Health Organization. http://www.who.int/governance/eb/who_constitution_en.pdf

WHO. (2002). Reducing Risks, promoting healthy life. In *The World Health Report*. World Health Organization. https://www.who.int/whr/2002/en/whr02_en.pdf?ua=1

Wiener, N. (1948). *Cybernetics, or control and communication in the animal and the machine*. MIT Press.

Wiener, N. (1950). *The human use of human beings: Cybernetics and society*. Houghton Mifflin Company.

Wiesing, U. (2008). The history of medical enhancement: From *Restitutio ad Integrum* to *Transformatio ad Optimum?*. In B. Gordijn & R. Chadwick (Eds.), *Medical enhancement and posthumanity* (pp. 9–24). Springer.

Williams, B. (1976). The Markopoulos case: Reflections on the tedium of immortality. In B. Williams (Ed.), *Problems of the self* (pp. 82–100). Cambridge University Press.

Wills, D. (2008). *Dorsality: Thinking back through technology and politics.* University of Minnesota Press.

Winance, M., Marcellini, A., & de Léséleuc, E. (2015). From repair to enhancement: The use of technical aids in the field of disability. In S. Bateman, J. Gayon, S. Allouche, J. Goffette, & M. Marzano (Eds.), *Inquiring into human enhancement: Interdisciplinary and international perspectives* (pp. 119–137). Palgrave Macmillan

Winner, L. (1989). *The whale and the reactor: A search for limits in an age of high technology.* The University of Chicago Press.

Wiseman, H. (2016). *The myth of the moral brain: The limits of moral enhancement*. MIT Press.

Witthøfft, N. L. (2011). The concept of nature and the enhancement technologies debate. In J. Savulescu, R. ter Meulen, & G. Kahane (Eds.), *Enhancing human capacities* (pp. 19–33). Wiley-Blackwell.

Wolbring, G. (2005). *The triangle of enhancement medicine, disabled people, and the concept of health: A new challenge for HTA, health research, and health policy*. Alberta Heritage Foundation for Medical Research.

Wolbring, G. (2006). The unenhanced underclass. In P. Miller & J. Wilsdon (Eds.), *Better humans? The politics of human enhancement and life extension* (pp. 122–128). DEMOS.

Wolbring, G. (2008). Why NBIC? Why Human performance enhancement? *Innovation: The European Journal of Social Science Research, 21*(1), 25–40.

Wolfe, C. (2010). *What is post-humanism?* University of Minnesota Press.

Wolpe, P. R. (2002). Treatment, enhancement, and the ethics of neurotherapeutics. *Brain and Cognition, 50*, 387–395.

Wudarczyk, O., Earp, B., Guastella, A., & Savulescu, J. (2013). Could intranasal oxytocin be used to enhance relationships? Research imperatives, clinical policy, and ethical considerations. *Current Opinion in Psychiatry, 26*(5), 474–484.

Young, S. (2006). *Designer evolution: A transhumanist manifesto.* Prometheus Books.

Zak, P. J. (2012). *The moral molecule: The new science of what makes us good or evil*. Transworld Publishers.

Zarpentine, C. (2013). 'The thorny and arduous path of moral progress': Moral psychology and moral enhancement. *Neuroethics, 6*(1), 141–153.
Zilsel, E. (2003). *The social origins of modern science*. Springer.
Zonneveld, L. (2008). *Reshaping the human condition: Exploring human enhancement*. Rathenau Institute.

The manufacturer's authorised representative in the EU is Springer Nature Customer Service Centre GmbH, Europaplatz 3, 69115 Heidelberg, Germany. If you have any concerns regarding our products, please contact ProductSafety@springernature.com

Printed and bound by CPI Group (UK) Ltd, Croydon, CR0 4YY

30/09/2026

02225701-0003